CONTENTS

ACKNOWLEDGEMENTS

This book is the result of a team effort. Many have contributed to its pages. The list is long. They know who they are and they know they have our thanks. However, special acknowledgement and thanks must go to the following; the authors who are mentioned at the top of each chapter, but particularly to Flight Lieutenant Andrew Layton, to whom fell the bulk of the research and writing; Squadron Leader John Willett who proof read and edited most of the chapters and Mr Ted Hirst who acted as mentor and guide for the younger authors unaware of the Base's history.

Readers will notice that not all events and activities have been recorded here. Likewise, not all the myriad of characters are mentioned either. The story of No 4 RAAF Hospital has already been written and West Sale is still to be covered. The authors and editors make no apologies for this, but wish to state that the omissions were not of our choosing. Space was the limiting factor.

A special thanks is extended to Air Force news, the RAAF Museum, Point Cook and RAAF Historical Section, Canberra. The photographs were professionally produced by CPE, Laverton and School of Photography, East Sale. Most photographs are RAAF official unless otherwise stated. Without this help, the information and illustrations could not have been collated in time.

As editor, I wish to thank the Commanding Officer RAAF Publications Unit and his team for producing and printing this book. The staff's professional approach and quality of work are self evident and finally, Air Commodore (Retd.) Deryck Kingwell, CBE, DSO for his support, encouragement and his very apt foreword.

Wing Commander Mark Lax
Editor
March 1993

2nd Edition with updating
chapter and index
April 1999

FOREWORD

BY

AIR COMMODORE D.W. KINGWELL, CBE, DSO RAAF (Retd.)

I was delighted, and felt honoured when I was invited to write this foreword because RAAF East Sale has always been dear to my heart and the celebrations for its fiftieth anniversary mean so much to me.

Even in the pre-war days Sale was associated with military and civil aviation. "Smithy" used to raise funds for his exploits by offering joyrides around the local area and Wapitis from RAAF Laverton regularly stopped over on bushfire patrols in the 1930's. On 22nd April 1943, RAAF Base East Sale was formally opened with the arrival of No. 1 Operational Training Unit, flying Beaufort and Hudson medium bombers.

The second World War saw rapid expansion of the RAAF, particularly in East Gippsland. Bairnsdale, West Sale and East Sale aerodromes all contributing to the war training effort. Over 3100 trainees, aircrew and ground crew, passed through the gates of East Sale and went on to help with Australia's war effort. Sadly, many were killed in flying training accidents because of inexperience and hasty training schedules.

After the war, Bairnsdale and West Sale became civilian aerodromes, but the RAAF kept East Sale as a major training base. Central Flying School, School of Air Navigation, Base Squadron and Maintenance Squadron being the current major units. I had the pleasure to be Officer Commanding post-war (1946-47) and saw the base grow as peacetime training tasks took over from the frantic programs of the war years.

Today, East Sale is one of the most modern Bases and one to be justifiably proud of its tradition and achievements. A massive rebuilding program has seen the demise of most of the old wartime huts with the best of modern facilities now available. It is also home to the RAAF's premier aerobatic team "The Roulettes", as well as about 850 service and civilian staff. The most recent addition to the Base was the re-establishment of No 32 Squadron in 1989, of which I was the first Commanding Officer when the unit formed in Port Moresby in February 1942. No. 32 Squadron is now a light transport squadron flying HS748 aircraft.

I congratulate the officers, men and women of RAAF East Sale on their fiftieth anniversary and commend this historical record to you. May the next fifty years be as good as those just passed. Those who follow inherit a modern base with a fine tradition of service to the RAAF and indeed, to the nation.

ALWAYS READY

A BRIEF HISTORY OF RAAF BASE EAST SALE

INTRODUCTION

by Flight Lieutenant Andrew Layton

When 19-year-old Jack King passed through the main gate of No 1 Operational Training Unit, RAAF Station East Sale in May 1945, he renewed a family presence at The Heart dating back several generations over 100 years. William Sawyer, formerly of Buckinghamshire in England, is a direct descendant of the King family and this journeyman brickmaker travelled to The Heart in 1844 to establish a brick works. The brick works would supply the materials to erect a homestead and station buildings for John Foster, a squatter who paid rent on the 29,440-acre property. The Port Phillip Government Gazette estimated The Heart run could graze 1,500 head of cattle or a flock of 12,000 sheep and by 1846, Sawyer had completed his labour on the station. Sawyer then sailed to Geelong from Port Albert and a century would pass before The Heart again engaged a relative of the Sawyer family.

The Heart is the title given to the district between the Latrobe and Avon Rivers and its origin has been attributed to various sources. Perhaps the most likely derivation of The Heart name appeared in a letter from Charles Tyers, Crown Lands Commissioner for the district of Gippsland, to the Lieutenant Governor of Victoria, Charles Latrobe. In the letter, dated July 1844, Tyers referred to the pastoral land between the rivers Avon, Thomson and Latrobe and claimed "it may be truly called the heart of Gippsland." Tyers wrote the final three words as a separate heading as if the name was generally accepted as a geographical point instead of a sentimental view.

By the 1890s, a state school, dance hall and church occupied part of The Heart and a nearby wharf on the Latrobe River handled the shipment of grains to Melbourne and Bairnsdale. When the government passed the Lands Act, cereal growing became the main farming activity on The Heart and dairying later replaced cereals as the primary produce after the implementation of irrigation into the area. After the Great War, English migrants emigrated to the region and toiled on the farms for the Australian landholders, earning each week a princely sum of about £1.

Nevertheless, beyond The Heart, in the rarefied world of international affairs, Australia remained a 'Dominion of the Mother Country'. On 3 September 1939, when Australians heard their Prime Minister, Robert Gordon Menzies, announce Great Britain's declaration of war against Germany, they learnt that Australia was then subsequently at war. Menzies address would eventually bring dramatic change to The Heart, directly affecting the district's landowners and townspeople of Sale.

Today, The Heart is better known as the site of RAAF Base East Sale and this book commemorates the Base's 50 golden years. In these few pages we have attempted to give the reader an insight into the development of the Base and more importantly, an appreciation of the people who have served their country within its gates. Not all who entered left. Many remain here, the result of tragic accidents and misadventure. This book is a memorial to them.

CHAPTER 1

CENTRAL FLYING SCHOOL

by Flight Lieutenant Andrew Layton

'A flying instructor should be an above average pilot of calm temperament with the ability to speak clearly and to be able to impart the correct techniques of flying training to the pupil.'

AIRCDRE 'Paddy' Heffernan (Ret)

When the main party of Central Flying School (CFS) moved to RAAF Station East Sale on 1 December 1947, it was the fifth and final move for the Unit in less than seven years. The Unit had reformed at Point Cook in 1940, but later moved to the provincial towns of Camden, Tamworth and Parkes in New South Wales before returning to its birthplace, Point Cook in 1944.

This is the 80th year since the establishment of CFS and its inception was due in part to the Aerial League of Australia which forwarded a proposal to the Military Board of the Department of Defence for an Australian Aviation Corps in late 1910. The Aerial League campaigned for the introduction of a military aviation school and aviation corps, within the control of the Defence Department. The Board deferred this plan on 17 January 1911 but the Minister for Defence, Senator Pearce, was instrumental in reversing the decision later that year. On 27 January 1912, the Commonwealth Gazette asked for 'Applications from Expert Mechanics and Aviators being appointed to the Defence Department.' The Department of Defence screened many applications and narrowed the selection to seven aviators. However, only two pilots were needed and after further deliberations, Henry Petre and Eric Harrison were chosen. Several mechanics also answered the Gazette notice and the Department of Defence recruited Messrs R. Chester, G. Fonteneau, C. Heath and A. Shortland.

On the 20th of September, the Minister for Defence approved the formation of a military flying school and enlistment would start in the new year. The Defence Estimates Committee allocated £14,031/16/- to the school and ordered two Royal Aircraft Factory B.E.2a tractor biplanes and two Deperdussin tractor monoplanes.

On a recommendation from Captain Oswald Watt, the Military Board allocated 640 acres for the School near Canberra's Royal Military College, Duntroon. As an aviation instructional staff member, Honorary Lieutenant Petre became an adviser for the establishment of the training school and after inspecting the site, denounced the region as unsuitable and that alternate sites at Langwarrin, Crib Point, Western Port and Altona near Melbourne were better. Honorary Lieutenants Petre and Harrison ultimately chose a property at Point Cook near Werribee which comprised 734 acres. This site became the initial location for CFS and the Aviation Corps and Australia's Federal Government formally announced the School's formation on 7 March 1913.

CFS Boxkite CFS3, the first military aircraft to fly in Australia - 1 March 1914

World Wars

Thirteen days after the Great War began, the inaugural Pilots Course started on the 17th of August 1914 and included four students who would form a Victorian Flight of the Australian Flying Corps. A further 11 courses were held between 1914-1918 which trained 152 pilots however, following the end of the Great War, the Governor General, Sir Ronald Ferguson, ordered the disbandment of CFS on 31 December 1919. On 31 November 1921, a military order renamed CFS as No. 1 Flying Training School and the School's aircraft were taken on charge by its successor.

Upon the outbreak of the Second World War, there was a huge demand for aircrew and the Empire Air Training Scheme evolved to train 11,000 aircrew in Australia each year for the duration of the conflict. In 1939, Australia had only 27 flying instructors and to fill this aircrew void, the Air Board reformed CFS on 29 April 1940 from 'C' and 'D' flights of No. 1 Flying Training School to conduct Service Flying Instructors Courses. The School lodged at Point Cook briefly until relocating to RAAF Station Camden on 14 May.

On 28 March 1942, less than two years after leaving Point Cook, the Air Board directed CFS to vacate Camden to make room for a United States Air Force formation. The Commanding Officer, Wing Commander D. Eayrs, directed his staff to inspect the New South Wales townships of Forbes, Young, Wagga Wagga, Mudgee and Corowa to determine their suitability as an alternative site for the School. Among these rural centres, the staff regarded Mudgee as the most appropriate site. On the 8th of April, Wing Commander Eayrs visited Tamworth and inspections of Kerang and Horsham in Victoria followed with Tamworth eventually chosen as the next home for CFS. In about a fortnight, a small road party drove to Tamworth to establish a camp and the School completed the move on 21 April of that year. The relocation of CFS to Tamworth led to an upheaval in the number of staff and aircraft previously at Camden. While at the Station, 111 officers and 593 airmen manned CFS with a fleet of 60 aircraft but by the end of April in Tamworth, the School's strength had dropped to 88 Officers, 430 airmen and 42 aircraft.

The School's stay at Tamworth was only temporary, as its training program expanded rendering the aerodrome's facilities inadequate. In 1943, concern had mounted regarding the availability of instructors at operational training units and CFS tabled a report addressing this matter for a conference at RAAF headquarters. The meeting decided to involve CFS in the supervision of instructor standards within operational training units and to do that effectively, the School required operational types of aircraft. At Tamworth, the School operated Avro Ansons, Wirraways and Airspeed Oxfords and the airfield's unsuitability for heavier aircraft led to the School's departure for RAAF Station Parkes on 18 January 1944. For almost eight months, the staff conducted Service Flying Instructor and Link Trainer Instructor courses, conversion courses, instrument flying and the tuition of pilots for operational training units.

Link Trainers of the type used at CFS

Between the 2nd and 5th of August the Commanding Officer, Group Captain R. Dupont, and the Chief Flying Instructor examined the resources at Point Cook and discussed with RAAF Headquarters, the feasibility of transferring the School from Parkes to Point Cook. These discussions resolved that the School could effectively use the facilities at Point Cook to enhance their training courses and this proposition would outweigh any advantages of remaining at Parkes. The School took about a month to move equipment and personnel by rail to Point Cook with the main party leaving Parkes on 18 September and detraining in Melbourne the following day.

The return of CFS to Point Cook was part of a general change in training throughout the RAAF and by the end of the Second World War, the School had expanded into an immense training organisation of 12 Elementary Flying Training Schools and eight Service Flying Training Schools. To fill the quotas of the Empire Air Training Scheme, the School had trained about 3,600 instructors who in turn taught more than 30,000 students.

The Post War Era

The declaration of peace also signalled a change of role for Point Cook. In 1947, the RAAF drafted a plan for officer and pilot training in the Interim Air Force and due to this impending plan, CFS transferred to East Sale between November and December to enable the formation and development of the RAAF College and Flying Training School at Point Cook.

CAC Wirraway A20-21 operated at CFS and relocated from Point Cook to East Sale when the Unit moved in Nov and Dec 1947.

On 18 November, a CFS advance party of one officer and seven airmen arrived at East Sale to ensure a smooth relocation for the main detachment due in a few weeks. Initially, the instructional staff occupied the existing buildings on the base until the flight huts were removed from Bairnsdale and flying training resumed the following day. The school was now under the command of RAAF Station East Sale and all its maintenance personnel had been posted to the Station's Maintenance Wing. On 26 November, CFS ferried seven Tiger Moths, three Oxfords a Lincoln and a Dakota from Point Cook to East Sale. In the second wave two days later, nine Wirraways, a Mustang and a Dakota flew to East Sale via Williamstown, Berwick and Moe.

After being attached to CFS for nine months, the first and longest Flying Instructors Course graduated from the School on the 28th of June 1948. The course was scheduled to take six months, but an intake of 20 students greatly outnumbered the eight staff instructors which included the Commanding Officer and Chief Flying Instructor. Furthermore, the low serviceability rate of the Oxfords prevented the completion of the Multi-Engine Night Flying and Instrument Flying phases of the course. However, the students passed the remainder of the syllabus which included Elementary Flying Instruction, Single-Engine Flying Instruction, Basic Instrument Instruction, Link Instruction and Multi-Engine Flying Instruction.

In addition to the Flying Instructors Course, the School was managing a Pilots Refresher Course, Fighter Conversion Course and the Flying Instructors Refresher Course. Among the numerous courses conducted by CFS, the Flying Instructors Course has undoubtedly become the most enduring with 114 of these courses held to date which have taught almost 1,400 pilots.

As another development of aviation in Australia, CFS started the Jet Instructors Course on the Dual Vampire in May 1953. In March the following year, the School added a new course to their instructional activities with the incorporation of an Advanced Flying Training Course under the control of the Jet Training Trials Flight. This was the RAAF's first attempt to use jet aircraft in the advanced flying stage of a pilots course and its trial was considered a success. The expansive learning program of CFS had earned it the reputation as the: 'university of flying with the sky as the classroom. The masters wore green flying suits instead of gowns and shiny flying helmets in place of mortar boards.'

After more than seven years since its inception, a Flying Instructors Course suffered its first fatality when Pilot Officer A. Pritchard of No. 12 course died in a Wirraway crash on the 13th of July 1954 at Pearsondale, three miles south of the West Sale aerodrome. Investigators of the Directorate of Flying Safety and the Directorate of Technical Services examined the accident. The Directorates had further inquiries to conduct as a Tiger Moth crashed two miles north of West Sale aerodrome only a week later. Lieutenant E. Clay and Pilot Officer D. Doel of the same course as the late Pilot Officer Pritchard were killed instantly.

No. 1 Flying Instructor's Course

Rear (L-R) S/L P. Metzler, F/L B. Brown, F/L N. Nixon, F/L R. White, F/L K. Andrews, F/L J. McKenzie
Centre (L-R) F/L N. Stuart, W/O H. Squires, F/L A. Kerr, F/L J. Cairns, W/O A. Dowrick, F/L K. McAtee, F/L C. Wright
Front (L-R) F/L K. Tasker, F/L L. McKay, F/L M. Holdsworth, S/L N. Quinn, F/L F. Schaaf, F/L R. Curtis, F/L A. McKenzie

In January 1957, CFS expanded its instructional role with the inclusion of a 'B' Flight and a short time later on the 12th of March, conducted its first Air Traffic Control Course. The School hosted the first three Air Traffic Control courses between 1946-47 but had relinquished this course to No. 1 Flying Training School upon moving to East Sale. Now in 1957, the School resumed its Air Traffic Control curriculum and provided basic and post graduate training for budding air traffic controllers. The RAAF formed 'C' Flight at CFS the following year to manage these courses and this lasted until the School of Air Traffic Control formed in 1981.

Accidents

After that tragic month of July 1954, Flying Instructor Courses continued incident free for about three years until coincidentally, a trio of students on No. 17 course experienced aircraft incidents in the same month On 1 May 1957, Flying Officer G. Hale was instructing Pilot Officer G. Talbot in a Vampire when they heard a 'loud bang' during a steep turn at 8,000 feet. The Vampire 'flick rolled' three times before the instructor regained control at 7,000 feet and successfully landed the aeroplane. A fortnight later, instructor Flight Lieutenant J. Horrocks (RAF) was flying with Lieutenant N. Lee and during a demonstration relight in a Vampire at 10,000 feet, the engine failed to relight and the instructor forced landed the aeroplane on the duty runway. The final incident occurred on the 16th of May when Flying Officer G. Hale and Pilot Officer M. Feiss found the Vampire's undercarriage would not lower and the speed brakes refused to close. The pilot manually lowered the undercarriage and applied 40 degrees flaps to safely land the aeroplane.

The introduction of the Winjeel to CFS on 10 October 1955 brought a change of trainer aircraft for the School. The Winjeel became the 27th different aircraft type which CFS had flown since the early days at Point Cook in 1913, and it would replace the Wirraway in 1957. On the 15th of October 1957, Wing Commander Vernon Sullivan led a formation of seven Wirraways to RAAF Station Tocumwal where the aircraft were disposed of from the RAAF.

By June 1959, a decade had passed since the first Flying Instructors Course graduated from CFS and in that month, the School expanded its training syllabus to include an Antarctic Course for RAAF pilots posted to the RAAF Antarctic Flight at Mawson. A Flying Officer pilot was a member of the first course and he was destined to renew his association with Gippsland in 1981 as Group Captain Graham Dyke, Officer Commanding East Sale.

Flight Lieutenant H. Wilson had just returned from the Antarctic Flight and was a member of No. 22 Flying Instructors Course when he and Pilot Officer J. Franks died in a Vampire crash at East Sale on 17 November 1959 The Vampire's engine failed on takeoff and Franks made a circuit of the airfield to touch down on the runway. He landed about 600 yards short of the runway however, and the aeroplane broke up as it hit an irrigation ditch. Several airmen approached the downed Vampire but were ordered back as its two 20-millimetre Hispano cannons were loaded for a firing practice sortie. Fortunately the crash had detonated only a few rounds of ammunition.

The completion of the Winjeel phase by No. 36 Flying Instructors Course marked the end of an era at CFS as they were thought to be the last students that would train on that aircraft. Winjeel is an aboriginal word that means 'young eagle' and the Commonwealth Aircraft Corporation built the aeroplane to replace the Tiger Moth and the Wirraway. No. 37 Flying Instructors Course would be the first course to train solely on jet aircraft. On 23 September 1966, No. 36 Flying Instructors Course farewelled the aircraft with a formation flypast of 10 Winjeels over Sale, East Sale and then Point Cook where they would serve as ab initio pilot trainers. Some of the frustrated students of No. 34 Flying Instructors Course did not enjoy flying the 'young eagle' and referred to the Winjeel as a 'stubborn little beast'.

CFI, Squadron Leader J. Jacobs, briefs the students of No. 36 FIC before a flypast

The Vampire had served CFS well

The Jets Arrive

The move to conduct 'all through' jet training for the RAAF would directly affect CFS and on the 27th of February 1965 the Commanding Officer, Wing Commander Tom Meldrum, accompanied an eight-man delegation overseas to select a new jet trainer. The team evaluated a number of aircraft and eventually chose the Italian Macchi-326 as a suitable replacement for both the Winjeel and the Vampire.

In 1967 CFS started its first 'all through' jet Flying Instructors Course in preparation for the introduction of the Macchi as the RAAFs new elementary and advanced flying trainer. The students of No 37, 38 and 39 Flying Instructors Courses posted to Pearce solely flew Vampires and from the next course onwards, the Macchi became the trainer aircraft for the jet instructors. Among the qualified jet flying instructors graduating at CFS in 1967 was Malaysia's first graduate of the school. Flight Lieutenant Ron Gophal Ramdas belonged to No. 38 Flying Instructors Course and after graduation, became the Commanding Officer of No. 9 Strike Fighter Squadron in the Royal Malaysian Air Force.

On 7 March 1968 the first Macchi accident at East Sale occurred when Flight Lieutenant Ward Rayner, on a staff continuation training sortie with aerobatics, ejected safely from the aeroplane after the aircraft's control column became jammed. Although Flight Lieutenant Rayner only received minor injuries to both legs, the aircraft crashed and disintegrated on impact on a farming property near East Sale. On the 15th of April 1969, a Macchi crashed on a grazing property near Perry Bridge, about seven miles from East Sale. The aeroplane hit the ground in a level attitude and burst into flames. Flight Lieutenant T. Spencer (RAF) and Flying Officer I. Cooper were on a low-level navigation exercise before the accident but died instantly. The aircraft sustained category five damage. Flight Lieutenant Spencer captained the aeroplane and instructed at CFS while Flying Officer Cooper was a student of No. 42 Flying Instructors Course.

Another Flying Instructor student died in a Macchi accident on the 22nd of July when the aircraft crashed one mile from the East Sale tower shortly after take off during touch and go circuits on a night flying exercise. Staff instructor Flight Lieutenant B. Newman ejected safely but Flying Officer H. Holsken did not eject and died when the aeroplane struck the ground and burst into flames. Flight Lieutenant Newman was admitted to the Gippsland Base Hospital suffering shoulder and spinal injuries and he was released from the hospital two days later.

GPCAPT Wilfred Lampe

A primary task of CFS has been the provision of refresher flying courses for the Australian Defence Force. In his East Sale history report for 1963, the Officer Commanding East Sale, Group Captain Wilfred Lampe, briefly explained the challenge confronting senior officers attached to CFS for refresher flying: 'The younger generation of flying instructors look with interest at having these old World War II veterans back with them to continue their flying careers. The interest and effort of the instructors is worthy of comment and so is the rather frustrated and tired feeling of the "aged" veterans during the first week of their conversions. On most occasions the time available by the officers for conversion is limited and to complete the syllabus a night flying phase of one to two periods at night must be added. Such a comment as "I haven't worked so hard for years" may be true but "I'm not as young as I used to be" is certainly true.'

Flight Lieutenant Ward Rayner ejected from his Macchi after the controls jammed and the aircraft subsequently crashed near the Base

The CFS Role

CFS would continue to offer refresher flying courses as required but a change in the role of the School was evident in the annual allocation of flying hours for the next financial year. The Commanding Officer, Wing Commander Herb Plenty, received advice that the two Canberras allocated to CFS would be withdrawn from the Base with no replacements. He questioned the reasons for the move: 'This poses the question as to the future position of CFS vis-a-vis its historical role as the mentor of pure flying techniques and procedures in the RAAF. During the past three years, the presence of Canberras at East Sale for use by CFS in maintaining twin-jet techniques and rating instructors on the type has enhanced the standing of CFS in fulfilling its function, particularly the annual recategorization of flying instructors in bomber squadrons and the examination and approval of Instrument Rating Examiners in these squadrons. CFS has been able to maintain its standing and prestige in annual examinations of bomber squadron Qualified Flying Instructors because CFS staff members were proficient in twin-jet techniques and procedures. With the withdrawal of twin-jet aircraft from CFS this proficiency will soon lapse and any attempt by CFS to continue this responsibility for examining and rating bomber squadron Qualified Flying Instructors and Instrument Rating Examiners will become effete and dangerous.

Historically, these considerations may well show themselves as the beginning of the time when the status of CFS began to decline to that of an instructor training unit only. A recent important pointer in this direction was established by the decision of the Department of Air that Qualified Flying Instructors and Instrument Rating Examiners of helicopter squadrons would be categorised and approved by the squadrons themselves.'

B Flight Crew Room - CFS 10 July 1963

'An exceptionally cold and wet winter this year has emphasised the inadequate heating and working conditions for flight crew rooms and lecture rooms.'

Wing Commander Herb Plenty

The Dakotas provided CFS with a twin-engine instructional role and in 1974 these aircraft trained the first five pilots of the Papua New Guinea Defence Force. Their graduation ceremony on 20 November at East Sale saw the Papua New Guinea Minister for Defence, Foreign Relations and Trade, Albert Maori Kiki, present the pilots with their wings. Kiki said his government appreciated the cooperation of the Australian Government and the RAAF for arranging the training of their pilots who reached the standard of first pilot day and night on Dakotas.

On 9 July 1975, Squadron Leader Brian Synnott was teaching Flight Lieutenant Mike Lavercombe single-engine flying in a Dakota when the right engine began to vibrate and within 40 seconds, the propeller separated from the engine and struck the right side of the fuselage. The propeller severed hydraulic and instrument lines and damaged the right-hand pilot's seat occupied by Squadron Leader Synnott. Neither pilot was hurt and Squadron Leader Synnott flew the aeroplane slowly back to East Sale and stopped it without brakes on the runway. A RAAF search party of 40 airmen later found the propeller on a farm about 22 kilometres south-east of East Sale.

Dakotas at Sale - 15 August 1951. The Base Met Section recorded the heaviest snowfall in the Sale district that morning

Of Plastic Parrots and UXBs

The arrival of CT-4 Airtrainers at East Sale on 30 August 1975 for No. 62 Flying Instructors Course ensured that the Winjeels era at East Sale was definitely finished. Winjeels had been recalled to CFS on several occasions due to a change of role for the School and when there were delays in the delivery of the CT-4 Airtrainers. The Commanding Officer, Wing Commander Arthur Barnes had more than 2,500 hours flying time in Winjeels and flew one Winjeel to Point Cook with the Officer Commanding East Sale, Group Captain K. Staib in the other aeroplane. The Winjeel first flew on the Base and across Gippsland skies in late 1956.

CFS became involved in an unusual incident on the 24th of May 1976 when a CT-4 was tasked to search for a missing 500-pound bomb released prematurely from an F-111 strike aircraft of No. 1 Squadron. The armour piercing bomb landed between Lake Coleman and Lake Wellington and contained 25 pounds of explosives which had not detonated. The Officer Commanding East Sale, Group Captain Hank Hurley said, 'it was the first time such a thing had happened here' and warned residents in the Dutson area of the potential danger. Ken Bramich, manager of the Dutson Downs sewerage farms, echoed the thoughts of other people in the area when he said, 'it could be buried in my backyard. It's not very reassuring when they say on the news that the release mechanism failed.' The bomb has not been found to date.

Unusual incidents continued to involve CFS when the staff instructors reported a bright light to the north of the Base during a night flying instructional sortie on the 2nd of February 1977. The light was reported to be a meteorite and an explosion was heard that evening in the Swifts Creek, Omeo area. Sale dairy farmer Col Ward watched the meteor's fall at 10 p.m. 'The planes from the RAAF Base were flying at the time and I thought one exploded. Then I saw a red fireball flanked by red streaks,' he said. The next morning, a large mud island appeared mysteriously near Pelican Point in Lake Victoria and local residents connected this event with the meteorite sightings. Scientific evidence showed there was no connection between the two happenings as geologists stated that the island's appearance was due to an underground eruption of natural gas.

While some people in East Gippsland were concerned with a bomb in their backyard, meteorite showers and a mud island in Lake Victoria, the staff at CFS continued their busy instructional program. The CT-4 was ably fulfilling the School's training role and a CT-4 at East Sale became the first of its type in the RAAF to fly 1,000 hours. On the 6th of December the Chief Flying Instructor, Squadron Leader Ron Biddell, captained the flight with Group Captain Hurley sitting alongside him.

The RAAF ordered 37 CT-4 Airtrainers in July 1972 at a cost of $2.827 million from New Zealand Aeroplane Industries and a Continental engine powered the aeroplane which flew at 170 miles an hour maximum speed. Throughout the late 1970s and 1980s, CFS conducted Macchi and CT-4 Flying Instructor Courses and both aircraft drew regular comparisons between the students. During one 'knocking session',the CT-4 was referred to as a 'black and yellow noise-to-drag converter' while the Macchi appeared to be an 'orange and white dog whistle'.

Nearly a year after a CT-4 recorded 1,000 hours flying time, another CT-4 crashed while carrying Santa Claus over a crowd of children at the annual base Christmas Treat on 13 December 1978. About 300 children watched Santa waving to them as the CT-4 circled the oval but on the second circuit the aeroplane clipped a 10-metre pine tree and fell to the ground and slid for about 30 metres before bursting into flames near the Airwomens' Block. CFS instructor Flight Lieutenant Paul Derbyshire (RAF) was unhurt in the accident but Santa, WOFF Tom Baker in disguise, received a fractured right leg and was taken to the Gippsland Base Hospital. WOFF Baker said, 'I was waving to the kids and next thing I knew we'd hit the ground. When I saw the flames, I thought I threw the beard and wig out, but I hadn't been able to find my beard. Paul grabbed me, my right boot was stuck in the fuselage and some other RAAF blokes had me out in no time.' Many children on the oval became quite upset as they thought Santa's death (in the crash) would stop them getting Christmas presents but the organisers calmed the children and announced that Santa had merely broken a leg. WOFF Baker was later transferred to hospital in Melbourne and then to No. 6 RAAF Hospital for physiotherapy. Santa said his injuries had not stopped the timely delivery of presents. 'You should see all the cards I got from the children who had seen me in the paper. I stopped counting at 137. There are too many to reply to individually and I have to get back to making toys,' he said. WOFF Baker had been Santa for about 16 years and was the last remaining World War II aircrew Senior Non-Commissioned Officer in the RAAF. A Squadron Leader at East Sale forecast that Santa would probably ride in a more traditional sled next Christmas but Santa chose a fire truck instead. The RAAF had flown CT-4 aircraft since 1975 and A19-057 was the first Airtrainer to be 'written off'.

Santa arrives by Winjeel - 10 December 1959

CFS is responsible for the maintenance of flying instructor standards in the RAAF, Army and the fixed-wing elements of the Royal Australian Navy, and has hosted many guests from overseas military services. The French Defence Attache, Colonel M. Bourghinaud, spent a week with `B` Flight flying the Macchi in February 1979 and he compared instructional techniques with the CFS staff. The French Air Attache, Lieutenant Colonel Francoise Drewi, visited the School in October 1975 and on one occasion he needed to telephone Canberra. As an authorisation for an official call, Flying Officer Bob Meares of No. 62 Flying Instructors Course told the switchboard operator that Lieutenant Colonel Drewi was the King of France. 'I am better than the King of France, I am a French fighter pilot,' Lieutenant Colonel Drewi said.

Recognition of Service

A burst of sunlight shot from a dull, grey sky as the Governor General, Sir Zelman Cowen, AC, GCMG, KStJ, QC stepped from his black Rolls Royce to present the Queen's Colours. Queen Elizabeth II had approved the award of the Queen's Colour to CFS in June 1977 in recognition of the Unit's past achievements. Only three Colours were held in the RAAF and the Colours presented at East Sale took 600 man hours to make at the Commonwealth Clothing Factory. The parade, a dual presentation with the School of Air Navigation (SAN), was held on the 28th of September 1978 and was the culmination of two weeks of rehearsals and several months of planning. A Regimental Sergeant Major from the Royal Military College, Duntroon and a Warrant Officer Disciplinary from RAAF Wagga assisted in the preparations. All members of CFS participated in the ceremony, either on parade, as part of the flypast team, as one of the

Roulettes, or assisting as an usherette. As Sir Zelman Cowen mounted the dais, there was a royal salute and a formation flypast of Macchi, Dakota, CT-4 and HS748 aircraft. During the presentation of the CFS Colour a formation of Macchi, Dakota, and CT-4 aircraft again flew over the parade and, following an advance in review accompanied by another flypast, the parade gave three cheers for Queen Elizabeth II. As the colours left the parade ground, the Roulettes presented an aerobatic display for the crowd of about 2,000. The ceremony was a distinguished occasion for the RAAF as it was the first time in Australian military history that the Queen's Colours were presented to two units on the same parade ground.

PNGDF Training

The Papua New Guinea Defence Force continued its instructional association with CFS in 1979 when Lieutenant David Inau became the first Papuan pilot to fly solo in a jet aircraft. He was a student of No. 72 Flying Instructors Course and graduated on 29 April as the first Qualified Flying Instructor from his country. Several Papuans attached to the School took time to accustom themselves with Australia while some of the Australian students needed time to adjust to their foreign course mates. An extract from the CFS line book highlights one problem that was encountered: 'After an obviously hard night of flying the guys piled into "Bombers Bus" to go home. "Bomber" promptly started the engine and waited. After three or four minutes of waiting he was asked what the delay was. "I`m waiting for Mark 'Corner' Storr" came the prompt reply from our erstwhile Navy aviator. It wasn`t till Mark smiled from the back seat that Bomber realised he had been there all the time.'

CFS and SAN Colours are paraded before the guard

The Demise of the Dak

The pioneer flying unit in Australia farewelled the RAAF's oldest aircraft with a Dakota formation over Gippsland on the 16th of March 1980. After 34 years with CFS, pilots of the School and from SAN, flew three Dakotas for the final time as the aircraft were being phased out of the RAAF. CFS instructor and Roulette pilot Flight Lieutenant Greg Walsh said he led the formation at 120 knots over Paynesville, Bairnsdale, Maffra, Sale and East Sale because he was the least experienced 'Gooney' pilot. The Gippsland Times reported the formation as a 'funeral procession' of three Dakotas with six Macchis, four CT 4s and three HS748 aircraft acting as the 'pall bearers'. The Gippsland Times described the scene on the base at the end of the formation: 'Then the wake at the Base Tavern... ground crew in mourning coats, top hats, the casket, a nose of a Dakota salvaged from Cyclone Tracey. Sad eyes painted on the windshield, tears trickling, a skeleton covered by an old RAAF great coat and the epitaph "lest we forget". At the rear a giant unserviceability tag signed "Nostradamus".'

CFS acquired its first Dakota in July 1945 and the School had operated a fleet of up to 11 Dakotas. The aircraft transported CFS staff on examining visits to military aerodromes across Australia and Papua New Guinea and supported the duties of Headquarters Support Command The Dakotas were also tasked for search and rescue, bushfire spotting and fishery surveillance. After the last Papuan students finished flying the Dakotas with CFS in late 1979, the RAAF directed the remaining 16 Dakotas be phased out of service. Flight Lieutenant Bob Graham, CFS instructor and organiser of the final flight, described the day's events as 'a glorious end to a flying era for an aeroplane which has survived many theatres of war.'

A Baby in the Family

After bidding goodbye to its oldest aeroplane, CFS, on the 11th of May 1981, supervised the birth of the RAAF's newest unit called the School of Air Traffic Control. Designated as 'C' Flight at CFS for the past 25 years, the new School would maintain a strong liaison with CFS through ground controlled approach and flying procedural requirements. For several years, air traffic control and flying courses at CFS had been open to Asian military services and the School enhanced its close ties with the Rebublic of Singapore Air Force by conducting an examining and liaison visit to No. 142 Squadron at Tengah between August and September. The Commanding Officer, Wing Commander Ron Biddell, led a group of seven Qualified Flying Instructors who flew a total of 80 Sorties on Hercules, Iroquois, Strikemaster, Marchetti, Skyvan, T 33a, Cessna 172, Hunter T755, F 5F and the TA 4A during their two week deployment.

It had been 70 years since the formation of CFS at Point Cook in March 1913 so, on the 9th of September 1983, the School celebrated its anniversary with a reunion at East Sale of past and present members. An afternoon function was held in the CFS Flight Buildings followed by aerobatic displays featuring the Roulettes, a solo Macchi and a CT 4. The reunion culminated in a dining-in of 91 officers held in the Officers' Mess with the Queen's Colour paraded to the musical accompaniment of bag pipes with the Chief of the Air Staff, AM David Evans, being the guest speaker. During the past 70 years of notable achievements, CFS had spent 36 years at East Sale and in the 1980s, the staff still occupied the same flight buildings the School manned in 1947. On the 25th of August 1986, CFS started to upgrade its accommodation when the headquarters in Battle Avenue and 'A' and 'B' Flights in Beaufort Avenue relocated to building 284 in Wirraway Street. The School would remain in building 284 for almost a year while a new building for CFS was under construction on the former site of its flight operations in Beaufort Avenue.

The Modern Era

On the 24th of November 1987, the Air Officer Commanding Support Command, AVM Alan Reed said, 'we have just witnessed history for the RAAF. The new arrival here today of the first PC9 signifies the beginning of a new era of RAAF pilot training.' AVM Reed was at East Sale for a ceremony to herald the delivery of the first PC9-A aircraft at a RAAF unit for operational duties. CFS received the first PC9 for instructor conversion training and to enable a comprehensive review of their flying training syllabus. The two PC9 aircraft on display for the ceremony were assembled in Switzerland and the other 65 aircraft on order from Pilatus would be built at Hawker de Havilland in Bankstown. A CT-4 and a Macchi escorted the two PC9 aircraft over the airfield and AVM Reed announced, 'we have decided to retain the CT-4 and the Macchi. We will have a three-aircraft training solution for our pilot courses into the future.' The additional PC9 aircraft would reduce jet hours on the Macchi from 150 to 60-80 and extend the life of that aeroplane and the CT-4 or another 10 years if required.

Initially, the PC9 aircraft at CFS displayed the standard RAAF training livery of orange and white before being repainted red and white in 1990-91

On the 3rd of August 1989, CFS experienced a Macchi crash for the third consecutive year when Flight Lieutenant Peter Szypula and Flying Officer Steve Jackson ejected after hearing radio warnings that smoke was streaming from their engine. The Macchi was conducting touch and go landings and takeoffs when the Commanding Officer, Wing Commander John Simmonds, flying nearby, and an air traffic controller warned the Macchi crew by radio. After ejecting, the two pilots parachuted to the ground and the aeroplane crashed into a Somerton Park Estate property about one kilometre north of the Base. The property's owner, Frank Simmonds, recounted his observations to an Age reporter: 'We (Frank and Dot Simmonds) heard two bangs, looked up and the jet's motor had cut out. Three seconds later it hit the ground and there was a huge ball of flame. The pilots were very lucky they missed the power lines and landed in a soft wet spot. We`re just thanking our lucky stars because we were standing in the same spot where it (the aeroplane) came down three minutes earlier.' At the time, 12 cows were about six metres from the crashed Macchi and they stampeded through an electric fence which resulted in the herd calving prematurely. He recalled that wreckage from the Roulettes crash fell in a neighbouring paddock last year. 'I think it`s time the RAAF looked elsewhere to do their aerobatics because it's getting too populated around here,' he said.

After flying over East Gippsland for the past 16 years, CFS farewelled the CT-4 Airtrainer on 10 December 1991 with an eight-aircraft flypast over Stratford, Maffra, Sale and the Base. The next morning the Commanding Officer, Wing Commander John Smith, led the CT4 fleet on a ferry flight to Point Cook where the aircraft were transferred to No. 1 Flying Training School.

Sid Overy's last flight 4 May 1983 - signoff before the traditional hose down!

About two years later, CFS experienced its first PC9 accident on the 5th of August 1991. The PC9 had completed nine circuits before departing the area for a TACAN approach to runway 27 and, after gaining a clearance from the air traffic control tower, the aeroplane crashed into the ground without an attempted ejection from either pilot or a radio transmission. The PC9 disintegrated on impact resulting in the death of staff instructor Squadron Leader Malcolm Cotterell and Malaysian student Captain Mohamed Tadjudin.

During 1992, No. 111 Flying Instructors Course became the first course to learn the new syllabus designed to produce a PC9 instructor capable of teaching ab initio students. A week longer than previous Flying Instructor Courses, the 17 week course included 105 flying hours in a PC9 and it assessed students on aircraft familiarisation, basic instructional techniques and applied instructional techniques. Flight Lieutenant Robyn Williams won the Wilson Trophy as the student who gained the highest marks in the ground school phase and she was the first female in the 45-year history of the Flying Instructor Course to graduate as a Qualified Flying Instructor. This award added to her past achievement of receiving the Hawker de Havilland Aircraft Australian Trophy as the Dux of the Course on No. 144 Pilots' Course.

Into the 1990's

By 1992, the ground school phase of the Flying Instructors Course had changed markedly since the early courses. Today, much time is devoted to instructional technique which culminates in a 50 minute assessed mass brief. In November 1952, No. 8 Flying Instructors Course learnt briefing skills from Professor Brown, Dean of the Faculty of Education at Melbourne University, when he spoke to the course on the 'art of lecturing and public speaking'. The teaching of instructional technique at CFS has since evolved and been refined over the past four decades. In August 1965, the course book for No. 34 Flying Instructors Course related their experience of instructional technique: 'To assist us in developing a technique of our own, we have started giving a series of 20 minute talks on any subject within the scope of the Flying Instructors Course syllabus, as at its current state of perpetual amendment. Some of the students are very gifted at this business, while others can't be heard for the rattling of kneecaps.' Most students prepared thoroughly for their briefs but Flight Lieutenant Kev MacKenzie on a later course, with perhaps an eye on the Wilson Trophy, carefully washed the blackboard on which he would conduct his 'pretty' brief and then blow dried the board with a fan heater.

The Flying Instructors Course allocates 70 per cent of the marks on flying assessment and there are 100 sorties for the student to fly. Of these sorties, about five were flown at night which some past students considered fairly trying. Flight Lieutenant Ian Smith of No. 72 Flying Instructors Course claimed confidently to his course, 'well, only six assessable rides to go.' 'You mean seven don't you,' replied a course mate. Flight Lieutenant Smith was adamant, 'hell no, you can't count the night flying readback,' which prompted a reply from all present, 'why not?' 'Well, we're all going to pass anyway, it would be too hard for them to organise night flying again.'

Upon the successful completion of the Instructor Categorisation, the Flying Instructor student graduates as a Qualified Flying Instructor D Category on the PC9. To many pilots, the Flying Instructors Course is perceived as the next challenge that awaits them after passing pilots' course and the joining instructions mailed to each student reinforces this perception: 'Upon graduation, you will know a great deal more about flying and your personal standards will have improved considerably. The title of "Qualified Flying Instructor" carries with it a level of responsibility that is not often shared by many junior officers and has many rewards, both personal and service wide.'

Aside from the Flying Instructors Course, CFS conducts flying refresher courses, tests Instrument Rating Examiners throughout the RAAF and assesses Air Force flying instructors. Through the flying instructional and examining roles, the School is responsible for the maintenance of pure flying standards in the RAAF. Several thousand personnel have passed through CFS as students or staff since the formation of the School in 1913 and in 80 years the School has operated almost 50 different types of aircraft.

Aircraft flown by CFS - Food for Britain Appeal - 25 May 1947

CHAPTER 2

No. 1 OPERATIONAL TRAINING UNIT

by Flight Lieutenant Andrew Layton

In past years, the Royal Australian Air Force (RAAF) had trained pilots as flying instructors at the Central Flying School while operational aircrews and instructors for the advanced training squadrons and service flying schools, completed their training in operational squadrons. The RAAF thought this system impaired the efficiency of operational squadrons as some aircrew did not gain a thorough grounding in operational training due to inadequate facilities in individual squadrons. This led to wide variations in instructional standards and training of these aviators was often drawn out over a long period of time. To compound this issue, there were no existing facilities in Australia to conduct operational training for aircrew to man the home defence squadrons or advanced training of staff pilots for Empire Air Training schools. In the northern hemisphere, operational training units were established in the United Kingdom, Canada and the Middle East and these units conducted advanced training for pilots, observers and air gunners prior to their posting to operational squadrons.

A New Flying Training Base

On 28 October 1941, the Chief of the Air Staff, Air Marshal Sir Charles Burnett (RAF), advised the Air Board of the necessity to form an operational training unit in Australia. He said the need for this type of unit had intensified since the 'recent deterioration in the war situation in the Pacific which had rendered it impracticable and unsound to use operational squadrons for training purposes.' He argued that if this Unit were stationed in Australia, it could direct the operational training of aircrew to serve in the home defence squadrons, train reinforcements heading overseas to RAAF home defence squadrons and instruct staff pilots for service at Empire Air Training Schools in Australia.

This unit would become No. 1 Operational Training Unit (No. 1 OTU) and commence training immediately at RAAF Nhill as a temporary location. Subject to Air Board approval, this unit would collocate with a General Reconnaissance School to a new site by 1 July 1942. On 27 November 1941, the Board of Business Administration concurred with the Minister for Air, Arthur Drakeford, that an operational training unit be established briefly alongside No. 2 Air Navigation School at Nhill and that £4,263 be set aside to pay for the associated buildings, services and armament ranges. Operational training started at Nhill on 22 December and accommodation would be available there until 1 July 1942 when No. 2 Air Navigation School was scheduled to double in size.

Drakeford advised the War Cabinet on 3 January 1942 that a site 4 miles east of Sale and 9 miles from No. 3 Bombing and Gunnery School had been chosen as a permanent location for No. 1 OTU, pending their approval. Additionally, he proposed that the General Reconnaissance School at RAAF Station Laverton reside on the same aerodrome as the operational training unit because the functions of both units were similar. He justified this move as a means to relieve the congestion at Laverton and it would equate with the Air Board's dispersal policy for the detachment of training units, wherever practicable, away from operational RAAF stations.

Beaufort A9-102 provided escort convoy patrols for No 1 OTU during 1943. The unit received the aeroplane in September 1942 and it had crashed at East Sale on 27 June 1943 when its undercarriage was retracted on landing.

The Air Board gauged that at East Sale, No. 1 OTU would train 60 Officers and 100 Sergeants at any one time in addition to 84 Officers under training at the General Reconnaissance School. Until new aircraft were available, the Air Board considered withdrawing a sufficient number of Hudsons, Wirraways and towing aircraft from the service squadrons and Empire Air Training Schools to allow training to start. Drakeford estimated this plan would cost about £260,000 based on costs associated with the development of bombing and gunnery schools at Evans Head in New South Wales and Port Pirie in South Australia.

Japan's surprise attack on the United States Pacific Fleet, berthed in Pearl Harbour on 7 December 1941 undoubtedly brought a sense of urgency to the construction of the East Sale aerodrome. Acting Air Vice Marshal Anderson requested approval to expend initially £33,800 for preparatory work on the aerodrome and Drakeford granted this request on 21 February 1942. In a previous submission to the War Cabinet, Drakeford issued a cost estimate of £260,000 toward the East Sale development and by 18 March he had revised that estimate with a building and engineering cost of £217,409. The unit would operate 60 military type aircraft and a provisional establishment totalled 1,053 personnel comprising 73 Officers, 144 Officers under training, 79 Sergeants, 100 Sergeants under training, and 559 men and 98 women of the Other Ranks.

On 29 March, Drakeford approved the acquisition of 1,186 acres, 3 roods and 26.4 perches of land for the site of No. 1 OTU. Acting upon advice from the Lands Valuation Committee, the Commonwealth planned to compensate £27,800 to the landholders of this site who owned all of the property except for five acres of Crown land. To secure immediate occupation of the site and to enable construction work to begin, possession of the land was taken under National Security (General) Regulations. On 3 May 1944, this recompense had increased to £29,015 when the Governor General of Australia signed a document for the compulsory acquisition of the land but negotiations ensued for another year until a payment of £31,225 was eventually finalised on 30 May 1945. Compensation was given to Stanton Overend, A. Andrews, Ellen Killeen, Joseph Glenn, Shire of Avon, the Victorian State Government and to the executors of the estate of the late Robert Cobains.

No. 1 OTU Arrives

When the War Cabinet approved East Sale as the site to accommodate No. 1 OTU in January 1942, the RAAF planned to operate Hudsons on a hard surfaced runway 1,300 yards by 50 yards in an east/west direction. However, by 11 May this plan had varied dramatically. Drakeford allocated an extra £57,650 to pay for an east/west runway 7,000 feet in length and a north-east/south-west 5,000 foot runway. The Civil Construction Corps started building the East Sale aerodrome in earnest on 20 April 1942 and trucks began transporting the first loads of about 217,000 cubic yards of gravel from quarries at Longford, Wurruk and Rosedale.

During the year of 1942, the encampment of No. 1 OTU was also subject to change. On 14 June, the Unit started to move its personnel and equipment to RAAF Bairnsdale after briefly lodging at Nhill and then mobilising to RAAF Station West Sale on 16 February. One week after No. 1 OTU completed its move to Bairnsdale, the Commanding Officer, Wing Commander Doug Candy, inspected the site at East Sale with the aerodrome's preplanner, Mr Godrey, of the Department of Home Security. He tabled various requirements for the new airfield indicating that about 150 medium bomber aircraft (Department of Aircraft Production Beauforts and Lockheed Hudsons) would be assigned to East Sale for operational training and to enable their dispersal off the aerodrome when not in use, he chose holding areas in the north-east, south, south-east and eastern sectors of the perimeter. This dispersal scheme accorded with Air Board policy and each area would comprise about 160 acres and accommodate 25 to 30 aircraft not less than 150 yards apart. These dispersal points would inhabit open grassed paddocks and gravelled taxiways with concrete hardstandings, 75 foot in diameter, were planned to ensure the aircraft

Group Captain Candy

did not bog during the wet winter months and to prevent dust problems and excessive engine wear in the drier summer period. This project gained an allocation of £37,500 and in May 1943, the unit took over two empty farm houses to use as flight line buildings for the squadrons dispersed afield of the main built-up sector of the aerodrome.

Since January 1942, the initial establishment figure of No. 1 OTU had been regularly revised and the RAAF now planned to house more than 2,300 personnel at East Sale. To accommodate these extra numbers, £53,114 was allocated on 29 June for more buildings and engineering services. However, the Chief of the Air Staff, Air Vice Marshal George Jones, expressed doubts as to the wisdom of placing 162 aircraft and this mass of people in one location. He thought it might be prudent to separate the unit between East Sale and Bairnsdale until the latter aerodrome accommodated operational squadrons.

In 1942, Air Vice Marshal Jones faced a predicament other than the Unit's establishment level and relief landing ground. He estimated that accommodation facilities at East Sale would be unavailable until 1 October so a decision was made to locate No. 1 OTU temporarily at Bairnsdale. By July, 1,630 service men and women manned the unit and after all the existing quarters, recreational buildings and 88 tents were filled to capacity at Bairnsdale aerodrome, 320 airmen were still left without a bed to sleep on. As an interim measure, the RAAF rented the Victory Guest House, and the Vic, Club, Orient, Albion and Main Hotels in Bairnsdale for £1,115 over 3 months. Overall, Drakeford allocated £5,229 to the temporary relocation of the unit which paid for the hotel rooms and the erection of temporary ablutions, latrines and tents situated out at the aerodrome.

Airmen from No 1 Operational Training Unit relaxing in a Bairnsdale Hotel in 1943. (Courtesy Maisey Lillicrapp)

Meanwhile, construction continued at East Sale and by October, there were sufficient facilities there to warrant a guard post. Leading Aircraftsman Norm Medew arrived on posting to No. 3 Bombing and Gunnery School on 24 October as an armament fitter but unbeknown to him, RAAF Station West Sale would not be his workplace just yet as he recalled in his book Up North in Forty Three: 'When we arrived we reported to the Service Police at the gate and were told to get into the back of a covered truck. In a few minutes, it drove away from the base and after passing through the town, pulled into what appeared to be a new town. We soon discovered this was the new East Sale air base and that we had been shanghaied to guard it. It appears that the Service Police at West Sale were responsible for security at East Sale but no guards had been provided. So they simply shanghaied airmen arriving at West Sale and held them incommunicado at East Sale until they were reported overdue. It was a simple matter then to find them. We were fed on a diet of steaks and potatoes and had no real complaints. Overall, about six airmen guarded the base on 2-hour shifts. Suddenly we were returned to No. 3 Bombing and Gunnery School and when we got to the Armament Section, Sergeant Cleary demanded to know where we had been. When we told him he was very angry, and declared he would put a stop to it.'

Operational training continued at Bairnsdale and the airmens' first intimation that a move to East Sale was imminent came in December when the Headquarters Orderly Room Sergeant pinned a typewritten page in the Mess notice cabinet announcing the impending relocation. On 20 April, No. 1 OTU commenced its move across East Gippsland to the emergent East Sale station. During the ensuing three days, Transport Section coordinated the immense exercise which involved 110 tenders that made 535 trips along the Princes Highway clocking up 50,000 miles. This move would have taken more time to accomplish without the support of No. 148 Australian General Transport Company which supplied 40 utilities. The unit's transfer involved the transportation of 2,411 personnel and 128 aircraft and was achieved without incident.

The East Sale airfield had already experienced a Beaufort incident on 18 March before the main party of No. 1 OTU arrived. Flight Lieutenant Adam Gillespie was flying a circuit of the airfield in A9-17 when an engine caught fire. The pupil pilot alongside him wanted to bail out but with the Beaufort at an altitude of only 1,500 feet, Flight Lieutenant Gillespie thought otherwise and physically restrained the nervous pupil in his seat while landing the aeroplane. A few pupils present for the next Beaufort course gathered to watch the scene and with the knowledge of numerous Beaufort accidents, were heard to say 'does that happen every time?'

A Fairey Battle flying over the Ninety Mile Beach. Known as the 'Fiery Bottle', No 1 OTU flew 14 of these aircraft when the station opened in April 1943.

Surprisingly, the move to East Sale did not hinder the flying program and training resumed in earnest the following day. After only 36 hours at the RAAF's newest station, the courses had already logged more than 200 flying hours. No. 1 OTU had a fleet of 25 Hudsons, 55 Beauforts, 35 Oxfords, 14 Fairey Battles, and one Tiger Moth and would train pilots, air observers and wireless/operator air gunners for multi-engine squadrons. The first part of each course embraced preliminary training and a 6-week conversion to the Hudson and Beaufort aircraft. While pilots were undergoing the conversion courses, air observers and wireless operator/air gunners received instruction in the Oxford and Anson aircraft. The trainees 'crewed up' by joint consent into crews which often remained intact for the duration of their operational tours. The instructors had served in operational squadrons and the sorties tested bombing and gunnery, shipping recognition and reconnaissance, navigation and searchlight evasion among other operational tasks.

The Air Board secured six armament ranges surrounding the East Sale Station for training purposes. At Longford, No 1 OTU operated a 400-yard target range which featured a model mobile target and power operated gun turrets at a cost of £22,260. (Courtesy Ron Appleton)

Not All Smooth Running

FSGT Ron Caffin

An overcast cold day greeted the majority of the unit's personnel to East Sale and Flight Sergeant Broughton noticed the aerodrome had reached an engineering stage of 'practical completion'. Nevertheless, the station experienced teething problems. The hot water system had not been connected and several personnel showered only twice that fortnight under the cold water in the shanty outdoor showers. Vehicles were driving along gravel roads offloading stores and personnel and new arrivals sought their living quarters. Tents temporarily housed airmen that could not be accommodated in the huts but on first impressions of the runway and buildings, the men thought that East Sale would become a permanent station for No. 1 OTU. Ron Caffin recalls the building that became his new home: 'It looked just like a suburban house with red tile roofs and weatherboard sides. However, they were merely unlined shells with no ceilings and gutters so that the keen winter blasts whipping down from the snow covered Baw Baw penetrated everywhere. Newspapers had to be stuffed around the top of the walls to try to stop some of the worst of the winds. Masonite partitions six feet high divided the houses into cubicles holding two trainees and the flimsy doors did nothing to help keep conversations private; in fact you could hear the person next door change his mind.'

Squadron Leader Bill Deane-Butcher slept wearing an overcoat under six blankets and spread sheets of newspapers between the top two blankets and on top of the mattress. As the Senior Medical Officer of No. 1 OTU, Squadron Leader Deane-Butcher arranged the issue of two extra blankets for everyone. An airman had tried to warm himself earlier by wrapping a brick, fresh out of a stove fire, into a blanket but it offered only momentary warmth before the brick burnt a hole through the material. The unlined huts not only let in cold winds but also invited possums into the rafters to bask in the relative warmth of the buildings. The huts inhabitants would shelter under raincoats while in bed to protect themselves from possum droppings when these marsupials ran across the rafters above them. Sometimes they would leap onto beds during the night and fall into boiling cooking pots in the Sergeants' Mess.

Mascots

'Snifter' the kangaroo, another member of the marsupial family, was a pet of the Women's Auxiliary Australian Air Force (WAAAF) contingent at the station. The WAAAFs brought Snifter from Bairnsdale and even though the kangaroo annoyingly pulled garments off clothes lines, Snifter had become a treasured mascot. The WAAAFs lived in huts within a fenced compound under armed guard in the evenings and they slept on stretches with hessian bags full of straw. During prolonged dry spells, some women contracted scabies when fresh straw was unavailable from the bedding store. The WAAAFs were assigned quarters according to their musterings and work schedules and the shift workers named their hut 'Boogie Woogie Bungalow'. The airmen huts in comparison displayed only a number and lacked the imaginativeness of the 'Beaufort Boudoir' title bestowed on a WAAAF's abode.

Keeping a Secret

The Works Board and the Camouflage Section of the Department of Home Security designed the layout of the aerodrome and the living quarters to resemble a country township. At a cost of £53,114, the quarters exterior closely matched residential houses with porches, chimneys, garden fencing and paths. The plan was unique to East Sale as it was the only RAAF station to accommodate its personnel in small houses instead of long rectangular, military style huts. This elaborate design however, escalated the aerodrome's overall cost which according to Allied Works Council estimates had reached £512,789 by early 1943.

Soon after arriving at East Sale, Group Captain Candy called a meeting of senior staff to order a tightening of security throughout the station. Bill Deane-Butcher remembers the discussion: 'He emphasised with some vigour the need to guard the perimeter of the base and to place well disciplined guards with orders to take no risks and, if necessary, to shoot on sight. Any questions? No one said a word. I tentatively asked if I could have a third medical officer in view of the stepping up of the shooting program. He was slightly speechless but in due course I got another medical officer.'

The expanded medical section unfortunately could not aid the Beaufort crew of four flying a night formation exercise on 28 April. Only 5 days after No. 1 OTU had resumed flying training, Beaufort A9-304 was seen to dive into Bass Strait three miles east of King Island and its crew have not been found since. This was the first accident for the unit at East Sale and its subsequent loss of life became a portent of incidents to follow.

The disappearance of Beaufort A9-304 would not be the last night-time incident to befall an operational training crew from the unit. Accounting for the establishment of about 130 aircraft, the Directorate of Training estimated the unit might total 2,000 hours monthly with an intensive night flying program. This expected rate of effort taxed the endurance of the aircrews and maintenance personnel and a mobile canteen fitted with thermos equipment offered a welcome catering service to their nightly training program. The canteen served suppers until 10 p.m. and as night flying increased, the passenger Chevrolet which carted the meals was replaced by a 2-ton Bedford Deluxe mobile canteen. The Salvation Army Womens' Home League bought the vehicle fitted with a radio, gramophone pick-up and extension speaker and it could serve 400 persons an hour. 'Mother' Hobbs was the face behind the counter who sold morning and afternoon tea for a penny to lines of airmen and senior officers throughout the station.

The acquisition of a recreational hut was the first of several recreational centres planned for East Sale. On 26 May, the Board of Business Administration endorsed an expenditure of £8,600 toward the construction of a gymnasium, airwomens' reading and writing hut, and a Sergeants' recreation room. The Chief of the Air Staff, Air Vice Marshal Jones, said it was of 'vital importance that facilities should be provided on the station to encourage and stimulate organised entertainments, athletics and musical and literary activities.'

The unit's instructional squadrons amused themselves between flights with betting games of Shove Ha'penny and Liar Dice. Each of the squadrons owned a Shove Ha'penny board and this hugely popular game was thought to have entered Australia via the hands of service men returning from the United Kingdom. Peter Creagh was a Flight Lieutenant instructor on Beauforts at East Sale and recalls the attributes needed to become a fine exponent of Liar Dice: 'Naturally the person with the best poker face and the least conscience prospered.'

Lockheed Hudsons were regarded among the trainee crews as a safer aeroplane than the 'Bogey Beaufort' but they were not immune from crashing. Almost 20 Hudsons had crashed between Sale and Bairnsdale by July 1943.

The personnel of No. 1 OTU were rostered to work 6-week shifts with a half-day off each week and the seventh week was designated as leave. When on duty, maintenance personnel worked until the aeroplane was serviceable regardless of the time and stood down only when it was ready to fly or the aeroplane had been classified as missing. Due to the demands imposed by the Empire Air Training Scheme, No. 1 OTU worked virtually 24 hours a day and throughout the unit and the RAAF, the station became known as 'Candy's Concentration Camp'. This nickname bore no resemblance to Group Captain Candy as a person who departed East Sale on 16 June on posting to No. 71 Wing Headquarters, leaving Wing Commander Keith Hennock as Temporary Commanding Officer.

Crashes, Prime Ministers and Bogeys

Sergeant Thaddeus Leach was a member of No. 10 Operational Training Course and on 12 June he departed East Sale with his crew in Hudson A16-32 to undertake a 4-hour training sortie. The aircraft lifted off into the darkness at 1:55 a.m. and Leach recalled it as: 'a scary flight. The white caps on the ocean surface looked very uninviting at night.' The Hudson flew over Paynesville, Flinders Island, King Island and the Hogan Group of Islands before returning to East Sale. As the sun rose that morning, Sergeant Leach and his crew landed at East Sale and headed to the Mess for breakfast, thankful to have seen the sortie through without incident. Hudson A16-32 which brought the crew safely home later denied another crew the glimpse of a Gippsland sunrise when they went missing over Bass Strait during a similar exercise on 6 July. The crew was on wireless silence and the Hudson disappeared into the darkness and was not seen again.

Edward Gough Whitlam, formerly of Knox Grammar School and future Prime Minister of Australia, came to East Sale and joined Sergeant Leach on the same inaugural Hudson training course at the new station. As a member of one of the 11 crews who finished the syllabus, the newly promoted Pilot Officer Whitlam proceeded on posting to an Embarkation Depot pending a further posting. Ron Whitwell remembers Whitlam on course and described the trainee as a bright man who was one of several aircrew susceptible to airsickness. As a safety measure, Whitlam often stowed a tennis racquet cover on board the Hudsons for the long sorties.

Pilot Officer Whitlam would survive the war and later gain admittance to the Federal Bar in 1947 as an early step to the prime ministership, but a crew on the corresponding Beaufort Course would get no chance to pursue a career after the war. On 11 July at 5 minutes to midnight, Flight Sergeant Ron Caffin departed East Sale in Beaufort A9-353 in formation with Beaufort A9-352 on a night training exercise. Ron Caffin remembers the flight: 'I was about an hour into the sortie and the chap flying pairs formation with me, reported engine trouble and decided to land at Currie Aerodrome on King Island. He didn't make it as he hit a hill on the approach and destroyed everyone in the aircraft. I didn't know anything about it until I was back on the ground because I was concentrating on my own fears to make sure I didn't go missing that night.' The Beaufort crashed 1.5 miles east of the aerodrome and apparently some of the wreckage is still among the rocks at the isolated crash site. Flight Sergeant Caffin's concern for his crew in A9-353 was well founded as the aeroplane crashed in August killing the crew and being ' converted to components'.

The loss of the crew in A9-352 was the third fatal Beaufort crash to befall No. 1 OTU at East Sale. Since operations started at Bairnsdale, the unit had experienced 47 Beaufort accidents and a fear of Beauforts was spreading among trainee aircrews. Reg Rechner was the Commanding

Officer of No. 1 Beaufort (Reserve) Squadron at East Sale and relates the story circulating the unit: 'When a new course came to East Sale, the blokes flying Hudsons almost had a smile on their face whereas the Beaufort blokes didn't due to a definite feeling of apprehension.'

On 15 July, Flight Lieutenant Rechner accompanied Squadron Leader Dick Maynard, Flight Lieutenant Adam Gillespie and Flight Lieutenant Arthur Sharp on a training exercise to No. 6 Service Flying Training School at RAAF Station Mallala. The school trained most Beaufort crews that came to No. 1 OTU and while at Mallala, Flight Lieutenant Rechner lectured to the students about the positive aspects of the Beaufort. He personally regarded the maligned Beaufort as a fine aeroplane and his enthusiasm for the 'Bogey Beaufort' earned him the title of the 'Beaufort Bogey Man'. On the return leg to East Sale, Flight Lieutenant Rechner presented a low-level flying display in the Beaufort to the station's trainees of No. 1 Elementary Flying Training School at RAAF Parafield. The school's Commanding Officer, Wing Commander Harry Plumridge, hauled Flight Lieutenant Rechner over the coals for the demonstration but the 'Beaufort Bogey Man' countered that he was trying to instil confidence in the Beaufort's flying capabilities among the students.

The Beaufort Bogey Man'

After the flying exercise to Mallala and Parafield, No. 1 OTU registered nine more aircraft accidents by 19 August and on that day, Group Captain Bill 'Bull' Garing arrived on posting to East Sale as the new Commanding Officer. Group Captain Garing acquired the nickname 'Bull' in 1931 for an ability to be heard giving instructions to a ground crew while circling above in an aeroplane at 1,500 feet. He had come to East Sale to halt the recurring aircraft accidents and to 'straighten things out' on the unit. Flight Lieutenant Gillespie had not yet met Group Captain Garing and he was in the showers one night chatting to a stranger about the relaxed environment prevalent among the instructional staff; there was no need to arrive on duty early and so forth and only later did the laid back instructor realise he had been speaking with the new Commanding Officer who came to 'straighten things out.'

GPCAPT 'Bull' Garing

Only 72 hours after Group Captain Garing arrived, No. 1 OTU suffered three devastating fatal Beaufort accidents in five days which provoked the Minister for Air, Arthur Drakeford, to ask the Air Staff for a review of all Beaufort accidents at training schools during the past six months. Once the report was written, Drakeford intended to brief Prime Minster John Curtin before informing the War Cabinet. Beauforts A9-175, A9-311 and A9-353 were all lost.

Four more aircrew were destined not to survive the month when they engaged in dive bombing exercises on a mobile splash target at Lake Wellington on 26 August. A target boat from the Marine Section towed the yellow, plywood and fabric construction along on a 400-foot steel cable and these exercises trained aircrew for future operational missions in New Guinea. Beaufort A9-353 dived from 2,500 feet at what reportedly was an "excessively sharp angle" but it failed to recover before hitting the water about 250 yards over the target. The crew of four died instantly and were entombed in the Beaufort, embedded in the mud 10 feet below the surface of Lake Wellington.

The seemingly never ending spate of aircraft accidents at Bairnsdale and East Sale led people to label the 25-nautical mile airspace between the RAAF stations as 'death valley'. The 'bogey Beaufort' tag given earlier to the aircraft had degenerated to the 'flying coffin' and unit personnel joked their spate of Beaufort accidents was exceeding the production rate of Beaufort aircraft at Fisherman's Bend.

Lake Wellington target launch

Four Beauforts sustained damage in unrelated accidents during a 3-week span after the final dive of Beaufort A9-353 into Lake Wellington. A decorated pathfinder veteran had been posted to No. 1 OTU on 16 August for instructional duties and he encountered his first Beaufort incident three weeks later. Beaufort A9-245 was just about to take off from East Sale when the port side tyre blew resulting in extensive damage to the airframe and slight damage to the airscrew. The Flight Lieutenant pilot was unhurt but his life would end suddenly in the unexplained crash of A9-303 on 13 September.

Crashes Investigated

Wing Commander Brett Hilder arrived at East Sale soon afterwards to preside over a court of inquiry into the crash of this aircraft and to investigate the causes of Beaufort crashes in the RAAF. Flying Officer Arthur Cambridge, a Beaufort instructor, assisted the inquiry which assessed the details of every downed Beaufort to identify any significant trends. After extensive research, Wing Commander Hilder suspected that the cause of the accident was the locking of the elevator trim tab and he recounted his flight tests in the Sun Herald: 'Cambridge and I flew every day or two in our attempts to get a Beaufort to 'bunt'. It took us about ten days to prove that it could be done by kicking the control column hard into the instrument panel.' Hilder had begun to close the inquiry when he read a report from Charles Taylor, a Victorian forensic scientist, concerning a 29 per cent level of carbon monoxide blood content thus rendering the instructor unconscious and often, the pupil. Taylor thought carbon monoxide gas was permeating Beaufort cockpits through a faulty ventilator in the nose when the aeroplane flew in cross winds with the engines operating a rich mixture. This hypothesis would later parallel Wing Commander Hilder's subsequent findings.

The Marine Section salvaging a Beaufort of No 1 OTU which had crashed near Paynesville. (Courtesy Harry Tate)

Wing Commander Hilder closed the inquiry now that he was satisfied with the cause of the crash to A9-303 but added 'that there were many other causes of the accidents as well.' The Advisory War Council convened on 10 November to discuss those causes and the Minister for Air, Arthur Drakeford, disclosed that 140 Beauforts had crashed between July 1942 to 31 October 1943 and among 21 unexplained accidents, existing evidence suggested a defective elevator control as the main cause. Aside from the carbon monoxide investigation and elevator control checks, the RAAF directed a statistical study to determine if a link existed between accidents and the maintenance effort per man on strength. Pilot fatigue was also being considered as a reason for the accidents but many pilots of No. 1 OTU suspected sabotage. After the meeting, the Department of Air grounded all Beauforts in the RAAF due to a defect in the aircraft's breeze control mechanism of the elevator trim tab circuit.

The Council's deliberations were greatly assisted by the evidence gathered days earlier from maintenance tests at East Sale and by the Commanding Officer of No. 5 Operational Training Unit, Wing Commander Brian 'Black Jack' Walker. On 5 November, Wing Commander Walker survived an incident in Beaufort A9-415 while flying between his unit at RAAF Station Tocumwal and Bankstown aerodrome. He forced landed the aeroplane and became the first and only pilot to land a Beaufort intact with a defective elevator trim tab. The Department of Air announced that Spindle Part No. Z.5083 was faulty due to 'an excessive undercut at the end of the threaded portion which results in the spindle being dangerously weak. Failure of the spindle renders the elevator control inoperative. In view of the nature of the elevator control in the Beaufort aircraft, it is necessary for pilots to use the trim tabs when entering and recovering from a dive. The standard finish of many of the spindles would indicate that it was regarded more in the accessory category that an item on which the safety of the aircraft might well depend.' The Department of Air concluded that defects in the mechanism might have caused a share of the

Beaufort accidents and the aircraft remained grounded until the Beaufort Division of the Department of Aircraft Production redesigned the defective spindle and manufactured a modified replacement.

October heralded another disastrous month for operational training at East Sale with eight aircraft accidents and six fatalities. On 5 October, Beaufort A9-33 attempted to gain altitude when about to land on the East Sale airfield but the port engine cut out and the aeroplane swung to port and banked sharply. The pilot failed to recover the Beaufort and the port wing struck the ground, cartwheeling the aeroplane as it burst into flames, killing the instructor and pupil pilot.

Two days later,just on dawn, Beaufort A9-409 crashed unaccountably into 10 fathoms of water about a mile and a half east of the Ninety Mile Beach near Woodside. The crew had just settled into a dead reckoning navigation exercises in a Beaufort that No. 1 OTU had acquired only three weeks earlier. The aeroplane was not salvaged from the depths of Bass Strait and the rescue party found only the body of the Sergeant wireless operator/air gunner among the crew of four men.

The Secretary of the Department of Air, Mr M.C. Langslow, had noted the unexplainable loss of life mounting among operational training units flying aircraft over the sea and he suggested the Army supply a pigeon service to complement the air sea rescue facilities at East Sale. The Department of the Army agreed to Langslow's proposal and deployed the 51st Australian Pigeon Loft to Gippsland on 30 November. The Australian Army Corps of Signals commanded the loft which comprised two other ranks members and 150 pigeons to serve No. 1 OTU which could have up to 30 aircraft over the water at any time. As a signals unit, the pigeon loft provided communications and training facilities at Seacombe on the shores of Lake Wellington until 30 November 1944 when the Unit was reassigned to New Guinea.

Women Contribute to the Base

Among the ranks of service personnel posted on strength were members of the Australian Women's Army Services who came to Sale for searchlight evasion exercises. About 40 women manned the detachment which operated searchlight batteries on properties in the Stratford district. The detachment's headquarters inhabited East Sale's satellite landing ground at East Stratford. No. 1 OTU tasked their trainee crews to fly over the Stratford area at night and attempt to evade the 200 million candle power searchlights which shone light up to 21,000 feet. The women lived an isolated existence in portable huts on the farming properties and fresh reserves of personnel replaced them every few months until the Army withdrew the detachment at the end of the Second World War.

Splash Bombing

Lake Wellington offered a convenient yet isolated location to conduct bombing exercises where Beaufort and Hudson crews practised against a defenceless target. By 5 January 1944, the Marine Section was stationed at Andrew Bay Landing and on that day, the ship's crew probably wished they had kept the armour plating the target boats originally were fitted with. Flight Sergeant Colin Hobson flew Beaufort A9-411 over Lake Wellington and banked in at a right angle to the target in a 40-knot crosswind and prepared to release a 25-pound practice bomb. He thought to himself: 'John (Flight Sergeant Crawford, the navigator) did not have enough drift angle applied but could do nothing about it as we were conducting low-level bombing using the bomb sight.

A few seconds after dropping the first bomb, the airwaves exploded with a string of expletives that would have made a bullocky blush. Next on the wireless transmission was a very curt voice which said, "the target is 200 yards behind the boat. The boat is not the target." To a question from our wireless operator, "how much did we miss you old man?" came the answer, "20 feet."'

A New Squadron

On 7 January, the Test and Instructional Training Squadron formed at East Sale under the command of Training Wing Headquarters. Coined 'TITS' by the menfolk on the station, the squadron consisted of the Test Flight, Fighter Flight and Instructional Flight. The Test Flight tested all the Beaufort and Hudson aircraft which came from Maintenance Wing after inspections on them were completed. Four Curtiss Kittyhawks and three pilots manned the Fighter Flight and taught fighter affiliation on operational training exercises. As the third flight in the squadron, Instructor Flight aimed to upgrade the standard of instructional flying and techniques of pilots posted from operational squadrons to operational training units. The RAAF created this flight to standardise instructor techniques and to maintain a high quality of tuition.

A Curtiss Kittyhawk of the Test and Instructional Training Squadron parked on the East Sale airfield. (Courtesy Reg Adkins)

Bush Fires Rage

On 21 January, 80 volunteers from East Sale backed the Stratford Bush Fire Brigade in burning fire breaks to shield two houses and they temporarily blocked an approaching fire. Three hours later, 250 reinforcements joined the gang of fire fighters to regain control of the blaze being fanned by strong 50-mile an hour winds. The Commanding Officer, Group Captain Bill Garing, cancelled all leave for the unit in expectation of worse fires which could start in Gippsland due to the hot daily temperatures and prevailing dry winds. On 6 February at midnight, the unit sent 150 men to protect the Willow Grove State Ash Forest near Moe and they prevented extensive fire damage to the region. Meanwhile, dense smoke blanketed Sale from fires south of the Latrobe River and in areas east of Stratford. The fires burnt about 5,000 acres of paddocks and many miles of fencing near Stratford as East Sale personnel, and the Maffra and Sale Bush Fire Brigades vainly tried to stop the advancing flames.

As the sun set behind a smoke filled sky, Meteorological Officer Pilot Officer Arthur Woolcock watched the fires surrounding the East Sale Station from the 'igloo' operations room. The meteorological office compiled regular weather forecasts for the squadron in the field and for Group Captain Garing who directed the deployments. There was no respite for No. 1 OTU as a fire which ignited in the Yallourn open-cut brown coal mine spread for half a mile. The unit deployed 270 men to the mine and about 500 servicemen tried to contain the blaze which cut electricity supplies and blacked out several Melbourne suburbs which delayed train services in the city.

The 'igloo' Operations Room was outside the front gate of East Sale to avoid destruction if enemy aircraft bombed the station. Visitors to the operations centre thought the 'igloo' resembled the plotting rooms seen on newsreels of the Battle of Britain. The 'igloo' closed on 6 August 1945. (Courtesy Nan Anderson)

The fires that ringed East Sale at the height of the emergency briefly hampered flying training which ironically had become safer than staying on the ground during the summer. No. 1 OTU had not recorded an aircrew fatality for about 5 months so it was surprising when the first pilot who died should end his life riding a bicycle rather than flying an aeroplane. Flying Officer Len Fuller had instructed at East Sale throughout the spate of aircraft accidents and survived a forced landing near a lake in Beaufort A9-282 on 16 May 1943. In September 1942, the then Sergeant Fuller served with No. 37 Squadron in the Royal Air Force and was awarded a Distinguished Flying Medal for his courage in a bombing mission over Palermo in Italy. His fine deeds as a military pilot followed his famous 'pick-a-back landing' in a paddock near Brocklesby, New South Wales on 29 September 1940. Leading Aircraftsman Fuller's Anson collided with another Anson from No. 2 Service Flying Training School and both aircraft wedged together 1,000 feet in the sky. After everyone else in both Ansons had bailed out, Leading Aircraftsman Fuller skilfully belly landed the connected Ansons, saving the RAAF about £40,000 and ensuring the aircraft could fly again. These incidents would endure in the annals of RAAF history but the young 25-year-old life of Len Fuller ended inexplicably at East Sale on 18 March 1944 when he struck a bus while peddling a bicycle.

Time Out

The autumn month of April brought cooler weather to Gippsland as the devastating bush fires of summer became a distant memory. The reading and writing rooms opened during the month with a library of 2,000 books for unit personnel to read in their spare moments, and fire places in both rooms comforted them during the bleak winter. The recreation hut regularly staged Thursday night dances to the musical accompaniment of the unit orchestra who entertained the personnel of West Sale, No. 4 RAAF Hospital and East Sale. On 9 April, the Airmen's Concert Party added a Musical Evening to the line up of social engagements in the WAAAF Recreation Hut. The musicians performed solo violin renditions of Beethoven's Minuet and Schubert's Cradlesong between solo pianoforte and piano accordion performances.

The cast and crew of the RAAF Concert Party, East Sale who presented Command Performance. (Courtesy John O'Toole)

Two days before the show, No. 1 OTU entered in the Metung Sailing Regatta and won two races and placed third in the Idle Along Class. The unit's sporting domination then extended to a service rowing regatta along the Mitchell River near Bairnsdale on 15 April. East Sale oarsmen won the Dahlsen Shield and the Gippsland Shield against crews from No. 4 RAAF Hospital, General Reconnaissance School and the Beaufort Division of the Department of Aircraft Production.

Disasters Resume

The happy experiences gathered by unit personnel during these social and sporting engagements dissolved on the morning of Anzac Day. They awoke to learn that Beaufort A9-417 had crashed five miles seaward of Lakes Entrance with its crew of four reported as missing. At 1:30 a.m., an observer of the Volunteer Air Observer Corps sighted a 'sudden illumination of the night sky' from the lookout tower on the Esplanade at Lakes Entrance. The Beaufort was flying a night exercise and fishermen near Lakes Entrance watched the lights of the aeroplane pass overhead and fly out to sea where they watched it plunge into Bass Strait. The rough seas and rain squalls prevented a search for the crew until dawn when the Lakes Entrance Harbourmaster, Fred Sutherland, and a group of fishermen guided Squadron Leader Reg Rechner and Flight Lieutenant Ray Kelly to the floating debris offshore. They recovered only one body.

No.5 Operational Training Unit (No. 5 OTU) had been conducting an almost identical Beaufort conversion course at RAAF Station Tocumwal to the initial conversion stage at East Sale. This similarity and the expanded role for No. 5 OTU persuaded the RAAF to divide the unit into a Beaufort Conversion Unit which transferred to East Sale under the command of the Commanding Officer, Group Captain John Ryland. The relocation of this training unit would reduce the duplication of instructors, maintenance personnel and supervisory staff necessary to manage Beaufort conversion courses. The release of the RAAF Development Program also indicated an expanded role for No. 5 OTU. On 12 July, about 550 personnel began the transfer from Tocumwal to East Sale which included trainee crews, maintenance personnel and instructional staff. This development swelled the personnel strength at East Sale to 2,576 by the end of July and increased again in November to 2,712 personnel.

GPCAPT John Ryland

Records Set

The infusion of a Beaufort Conversion Unit at East Sale produced a series of record operating figures in August for No. 1 OTU. During the month, the unit flew 2,479 flights totalling 3,868 hours with a fleet of 132 aircraft which taught 221 pupils. This busy yet safe program encouraged the Commanding Officer, Group Captain John Ryland, to mention the 'very low accident rate of the one time "Beaufort Bogey". This theory of jinx is exploded for all time, and when it is considered that one only fatal accident has occurred in 24,424 Beaufort hours, it will be agreed that Beauforts were indeed misjudged.' Earlier faults blamed on the elevator trim tab mechanism, pilot error and carbon monoxide in the cockpit were now replaced by the unwelcome influence of the 'Lake Gremlin'. The 'Lake Gremlin' was a fictitious character that allegedly dragged aircraft into Lake Wellington inflicting damage to airscrews and wingtips and occasionally 'writing off' Beauforts. The blame given to the gremlin by maintenance personnel, instructional staff and training crews seem justified in their minds after many aircraft had been damaged or destroyed while flying over Lake Wellington.

The officers of No 1 OTU when GPCAPT John Ryland was the CO - late 1944. (Courtesy Reg Eddy)

East Sale's recent safe aircraft record ended on 16 September when Beaufort A9-7 crashed one mile north-east of Forge Creek near Bairnsdale during a low-level formation exercise. General Reconnaissance School at RAAF Bairnsdale despatched Avro Ansons to search the area and the Forge Creek Postmaster, Mr Clarke and daughter Mrs Delaney, guided an ambulance to the crash site on a lightly timbered paddock. The Beaufort disintegrated in the crash killing the Flight Lieutenant pilot and seriously injuring the Flight Sergeant wireless operator /air gunner. Moments before the accident, the residents of Forge Creek had watched three aircraft fly over the town and soon afterwards one engine on Beaufort A9-7 failed and the aeroplane stalled on the other engine.

Low-level formation flying was an element of the Beaufort Conversion Course which also directed anti-submarine exercises off the Victorian coast. No. 1 OTU tasked operational patrols to safeguard the shipping lanes and in April 1944, four Beaufort aircraft from the unit deployed to Mount Gambier to provide a two-day escort patrol for the Dutch transport ship Van Ruys. These patrols proceeded without incident as did similar missions in July for a troop ship sailing across Bass Strait. Indeed, convoy patrols which actually sighted enemy submarines were rare and Air Vice Marshal William Bostock, Chief of Staff to the Commander - Allied Air Forces - South West Pacific Area, ordered a cutback of RAAF anti-submarine patrols along the Australian seaboard on 12 February 1944. There were further reductions in convoy duties two months later but No. 1 OTU was still given the occasional escort duty.

This enemy inactivity changed in September when military intelligence reported that German submarines might soon appear off the Australian coast. The reports suggested that shipping lanes in the Eastern Area and Southern Area were within range of the German navy. Southern Area commanded No. 1 OTU and alerted Beaufort aircraft from the Unit to hunt a

German submarine which shelled the Greek motor boat Illosis on 9 December. Korvettenkapitan Heinrich Timm skippered U.862 which accidentally confronted the Illosis on a misty summer afternoon near Lacepede Bay, Kingston in South Australia. The U-Boat fired four shells from its quarter deck gun which flew wide and astern of the launch and then dived soon afterwards. The people on board the Illosis failed to identify the raiders and it was not until the end of the war that a captured German Naval Staff log book verified the incident.

On 8 December, the day before Korvettenkapitan Heinrich Timm evaded the Beauforts of No. 1 OTU, Group Captain John Ryland opened the East Sale swimming pool which soon became known as 'Lake Ryland'. The pool measured about 200 x 100 yards with a depth of almost 40 feet. Its unlined bottom muddied the water but the diving platforms, pontoons and swimming lanes marked with rope and cork flotation lines, enticed many bathers for a dip in 'Lake Ryland' during the hot summer of 1944-45. Swimming and diving contests followed the official address and rated a column in the Gippsland Times.

The opening of 'Lake Ryland' was one of the last official duties for Group Captain Ryland at East Sale as he handed over command of No. 1 OTU to Wing Commander Geoff Nicoll on 13 December. That evening, the No. 1 OTU Revue Company staged Out of the Blue to the new Commanding Officer and civilians and service people in Gippsland. The stage curtains were drawn back to reveal a mock pirate ship and the opening scene entitled Under the Jolly Roger presented four songs to start the three and a half hour show of comical and musical performances. The Revue Company had become renowned for their professionalism and their reputation filled benches in the Maffra Mechanic's Hall on 18 December and the Prince Regent Theatre in Bairnsdale the following evening. Out of the Blue displayed performances usually seen in Melbourne theatre houses because No. 1 OTU had luckily drawn together a band of gifted entertainers who performed professionally in city venues before they enlisted in the RAAF. Several of these entertainers had performed for the famed J.C. Williamson Company in Melbourne which, with the Tivoli Theatre, supplied authentic costumes for the revue performers. Lionel Corrick arranged and orchestrated the musical score for Out of the Blue and brought to the show fine musical skills that were honed on stage in Melbourne's Capital Theatre.

The final year of the Second World War and for No. 1 OTU started on a sombre note for the aircrew stationed at East Sale. Twelve days into the new year of 1945, Beaufort A9-426 departed East Sale on a night flying exercise and for reasons unknown, the aeroplane failed to return its crew of four to the station. Today, these men are still classified as missing.

FLTLT Bob Bennett forced landed Beaufort A9-146 on 23 November 1946 after he discovered the starboard undercarriage leg would not lower. The crew of four were unhurt but the Beaufort was destined for destruction. On 7 February 1945, the aeroplane crashed 5 miles east of the station on a night flying exercise, killing the crew of four men. (Courtesy Kevin Fahey)

Four days after Beaufort A9-426 went missing, the Test and Instructional Training Squadron lost its first pilot and Kittyhawk. The pilot was flying a fighter affiliation exercise in Kittyhawk A29-71 when his error of judgement crashed the aeroplane into Lake Reeve near Seaspray. The Kittyhawk remained undisturbed in Lake Reeve until 1951 when the RAAF salvaged the aeroplane's cockpit components and browning machine guns loaded with ammunition. In April 1992, the water level of the lake shallowed to such an extent that it exposed the wreck of the Kittyhawk to an inquisitive public unaware of its history at East Sale. A part of the fuselage and propeller lay corroded and embedded in mud and the wreck bore little resemblance to a fighter bomber which graced the skies over Gippsland almost 50 years ago. The Lake has since reclaimed her prize.

The RAAF attributed the crash of Kittyhawk A29-71 to pilot error but the loss of Ventura A59-9 with six men on board was due to a dense fog which rolled in about mid-afternoon on 23 February 1945. Four radar mechanics accompanied the Flying Officer pilot and Warrant Officer wireless operator air gunner on board the Ventura to conduct a radar test flight and their crashed aeroplane was found the next morning near East Sale. The incidence of aircraft accidents due to fog in the area was however, surprisingly light but this weather phenomenon continues to challenge aircrew today as East Sale is recognised as the second foggiest aerodrome in Australia.

The notorious Gippsland fog returned in force the next day to wreak havoc with the night flying program of No. 1 OTU and the General Reconnaissance School at Bairnsdale. Sixteen Avro Anson aircraft from the school were conducting night flying exercises and became lost between Morwell and Bairnsdale. Five Ansons crashed due to the weather and among the 60 aircrew airborne that evening, three men died.

On 6 April, Wing Commander Geoff Nicoll handed over command of No. 1 OTU to Group Captain Geoff Hartnell and the outgoing Commanding Officer departed in triumph as coach of the eight that won the Victorian Open Championship on the Yarra River in Melbourne. The crew fielded three Kings Cup oarsmen among the junior officers and airmen which represented the Sale Rowing Club. Chaplain Flight Lieutenant Gillam 'Pat' Wood rowed in the victorious crew and he had previously coached the eight in his spare time. On 25 November 1944, Flight Lieutenant Wood rowed in place of a man stricken with malaria and the chaplain helped the eight to victory across two 1-mile courses at Henley in Melbourne. He then boarded a night train bound for Sale and rose early Sunday morning to deliver a sermon and round off a hectic weekend.

Chaplains Role

The chaplains associated with No. 1 OTU had actively fostered sporting and social events at East Sale during the past two years. They supervised a fellowship meeting in the recreation hut every Friday evening, arranged debates on topical issues and led hikes across the scenic Gippsland bushland on standown days. A chaplain chaired the Unit Welfare Committee and acted as librarian in the Officers' Mess until February 1945. The unit revue company owed some of its success to the chaplains of East Sale and the musically gifted minister Flight Lieutenant Fred Angus, who applied his talents to stage these productions. For a training unit which had suffered an appalling number of fatalities, the chaplains of No. 1 OTU provided an indispensable service to the grieving relatives and friends of the deceased servicemen. These ministers kept close ties with the clergy in Sale who allowed the RAAF to use their churches for service funerals before the bodies were laid to rest in the Sale cemetery. By the end of the Second World War about 50 service graves dotted the Sale cemetery.

Innumerable aircraft accidents had made service funerals depressingly common in Sale. (Courtesy Ron Appleton)

The chaplains moral support was needed again on 4 June when Beaufort A9-228 crashed near Tallangatta. The Beaufort took off from Mount Gambier at sunset on a long range reconnaissance and night astro navigation exercise and neither the crew of four nor the aeroplane were ever seen again despite widespread land and air searches.

Fortunately for the crew on board Avro Anson W1991 of General Reconnaissance School, a search and rescue launch picked them off Snake Island near Corner Inlet after a night sortie ended unexpectedly in Bass Strait. The crew had lost direction in a dense fog on 18 June and circled the coast until an empty fuel tank forced the pilot to dump the Anson into shallow water near the southern end of Snake Island. To compound their plight, the crew inadvertently discarded the emergency receiver among other items to empty the Anson as a means to stay airborne until the fog lifted. At sunrise, the Volunteer Air Observer Corps and the Yarram Police searched in vain along the Ninety Mile Beach near Woodside where the aeroplane was thought to have crashed. A search aircraft found the missing aviators soon afterwards on Snake Island and the Marine Section launch brought the men to Welshpool suffering minor injuries and exposure.

Andrew Bay Landing on the shores of Lake Wellington was home for the Marine Section of No. 1 OTU and those men in the section not listed as duty crew on 21-22 June could take the first 'long' two-day weekend for several years. Since Germany unconditionally surrendered to Britain, the United States of America and Russia on 7 May and with an Allied victory seemingly assured in the Pacific; the unit's work rate had lessened. The personnel of No. 1 OTU were now free to enjoy two days leave per week. More than 900 personnel took the train to Melbourne that weekend and the Sale Bus Service and Transport Section provided nine buses to carry them the short distance to the Sale Railway Station.

The declining training effort of No. 1 OTU did not lead to a drop in aircraft accidents between July and August. On 2 July, Beaufort A9-587 crashed into a paddock several hundred yards beyond the duty pilots' tower. The tip of the Beaufort's airscrew had snapped off and flew into the cockpit and the two crew members died when the aeroplane hit the ground and caught fire. A further accident occurred on 19 July when after a fire in the port engine, Beaufort A9-101 crash landed at Letts Beach. One crew member, Flying Officer Bruce Hinge, was only slightly injured but the other member, Flying Officer Ossie Morgan sustained skull fractures and his flying suit and hair caught alight in the resultant fire. Flying Officer Morgan was admitted to No. 4 RAAF Hospital and the medical staff worked all night to stabilise his condition. He was later transferred to No. 6 RAAF Hospital, Heidelberg where a plastic surgeon operated for several hours to repair the damaged skin.

No. 6 Operational Training Unit was operating the same Beaufort in August 1943 when the aeroplane lost the starboard airscrew after it skimmed the ocean during a simulated torpedo run on a shipping column south of Wollongong. Flight Sergeant Bob Roberts skilfully flew the damaged Beaufort back to RAAF Station Nowra where the aircraft was repaired and later transferred to East Sale. Flight Sergeant Roberts also went to No. 1 OTU but like Beaufort A9-101; his time at the unit expired suddenly. The promoted Pilot Officer Roberts died in Beaufort A9-71 as it crashed on take off and was destroyed in the ensuing fire on 29 December 1944. Only the airscrew of A9-101 exists today and it hangs in the Lady Denman Historical Complex at Huskisson, south of Nowra.

When Beaufort A9-143 crashed on Waddy Island about three and a half miles south-east of Paynesville, it was the fifth aircraft accident for No. 1 OTU in July 1945. The crew of three were killed instantly on impact and the grieving parents of the pupil pilot visited Paynesville a year later to see where their son had died, only two weeks before the war ended. The final aircraft accident to befall a crew of No. 1 OTU killed three members of the station's last Beaufort Transport Course on 27 August. Beaufort A9-305 crashed at Puralka, 18 miles east of Mount Gambier while undertaking a training exercise. In about three and a half years, No. 1 OTU had suffered 147 aircraft accidents in southern Australia and New Guinea which left 131 aircrew dead or presumed missing during the unit's intense program of training exercises and operational missions.

Beaufort A9-562 was taken on charge by No 1 OTU on 15 January 1945 and operated at the station until 10 August when it crashed about a mile north of East Sale. (Courtesy Jim Prendergast)

The War Ends

On 15 August, No. 1 OTU received confirmation of Japan's surrender to the Allied forces and the unit stood down immediately for four days. About 400 personnel who attended the all ranks dance in the station gymnasium the previous evening heard the news at 10 p.m. Flight Lieutenant Roy Lugg was on stage when he heard of the Japanese surrender and announced, 'ladies and gentlemen, I have been advised that Japan has surrendered.' Silence greeted the news. Then, several WAAAFs began to cry and emotional scenes surfaced for 10 minutes until a sense of relief overcame the crowd and cheering broke out. The 12-piece orchestra then resumed playing and the dance continued. The next day, No. 1 OTU held a Thanksgiving Service in the gymnasium and Chaplains Squadron Leader John Brown and Flight Lieutenant Gillam Wood conducted the service assisted by the Temporary

FLTLT Roy Lugg by Athol Snook

Commanding Officer, Wing Commander James Hepburn, and the unit Adjutant Flight Lieutenant William Treyvaud. Owing to his abstinence from alcohol, Flight Lieutenant Lugg was appointed orderly officer for the victory and peace day. He discreetly stowed the RAAF Ensign from souvenir hunters after lowering it from the masthead at sunset but a thief located the flag before sunrise.

The public of Sale celebrated the occasion with a fancy dress procession from the post office to the Sale Cricket Ground. Sports and novelty events were contested at the ground that afternoon and in the evening a bonfire at Lake Guthridge attracted a large crowd who sang songs and rejoiced at the resumption of peace.

IN GIPPSLAND HILLS

Hid' high on Gippsland mountain side
torn, twisted metal, scattered wide
Beneath the forest's canopy
where questing eyes shall never see.

Where timbers tall and fern glades are
lie crumpled wings with shattered spar.
Air-screws bent on furrowed ground,
a wartime relic never found.

Some moulding forms in Airforce blue
that once were men like me and you.
No hostile bullet ever sped
and yet those boys are just as dead.

Grim sequel where the eagles soar
of stealthy hand of sabateur.
Lost mates of ours we fondly knew
from flying 1.O.T.U.

CHAPTER 3

AIR ARMAMENT SCHOOL

by Flight Lieutenant Andrew Layton

'Doceo ut Pugnent' - "I teach, that they may fight'

- Motto of Air Armament School

The disbandment of Air Gunnery School at RAAF Station West Sale on 27 November 1945 brought a close to the School in name only as a nucleus detachment mobilised to RAAF Station East Sale and became the Air Armament School. The creation of this unit was directly attributable to a reformation of the post-war Air Force and in line with the establishment of RAAF Station East Sale as a separate Formation. The School attained official status in January 1946 and it combined the roles previously executed by the Air Gunnery School, Central Gunnery School at RAAF Cressy and Air Armament and Gas School at RAAF Nhill.

This trio of units transferred stores and equipment to the Air Armament School and in February, six officers arrived on posting at East Sale to prepare for the School's first course in May. The School's staff received advice on armament instruction from a five-man delegation from the Empire Air Armament School who visited East Sale on the 22nd of May to lecture on armament equipment and training in the Royal Air Force. The team flew the Lancaster 'Thor' to East Sale and aircrew at the station were shown its armament equipment, automatic gun laying turret, stabilised automatic bomb sight and radar system. The RAAF Director of Armament, Group Captain Ivor Lightfoot and Wing Commander Garth Miles from the Directorate of Training accompanied the RAF team and inspected the School at the end of the tour. During the School's next decade of training, staff primarily taught aircrew to use airborne armament systems while collating a comprehensive reference library to maintain a thorough knowledge of contemporary weaponry.

The first Pilot Gunnery Instructors Course of the Air Armament School began on the 26th of October 1946 and though the course gave the instructors a specific commitment, the scarce furnishings and equipment hindered the six pilots in training. Several air gunnery ranges were near the East Sale Station but the School commanded just three Mustang aircraft and all were inadequately armed for the six week course.

The PAI Course beside a School Mustang at East Sale - 13 February 1950

In 1947 the Air Armament School augmented its instructional curriculum with a Gunnery Leaders Course and an Ammunition and Explosives Course. The flying staff now operated four Mustangs named 'Mongili' I,II,III and IV, the name being derived from a Gippsland Aboriginal dialect which meant 'warspear'. The Mustangs were emblazoned with a war spear but the Unit's heraldic emblem and motto remained in a draughtsmen's office out of public view and awaiting further design work.

Just over two years had elapsed since a liaison tour by the Empire Air Armament School and on the 27th of May 1948, the School's representatives returned to East Sale. The team landed in 'Thor' II and discussed armament issues with their RAAF counterparts. The second tour highlighted the slow progress the Air Armament School had made during the past two years. Though the bomb and bomb carrier demonstration room, gun room and turret room, and gunsighting section were nearing completion, 'the work on bombing and gunnery ranges was so slow as to seem almost hopeless,' the Commanding Officer, Squadron Leader Edgar Slater said.

In August, Air Armament School graduated its initial Aircraft Recognition Instructors Course and subsequent courses attracted Navy and Army members to the School. The syllabus enabled aircrew to quickly and accurately identify aircraft, ships and armoured vehicles. In conjunction with a course to identify modern instruments of war, the School also managed a section which researched and instructed on the use of nuclear and guided weapons.

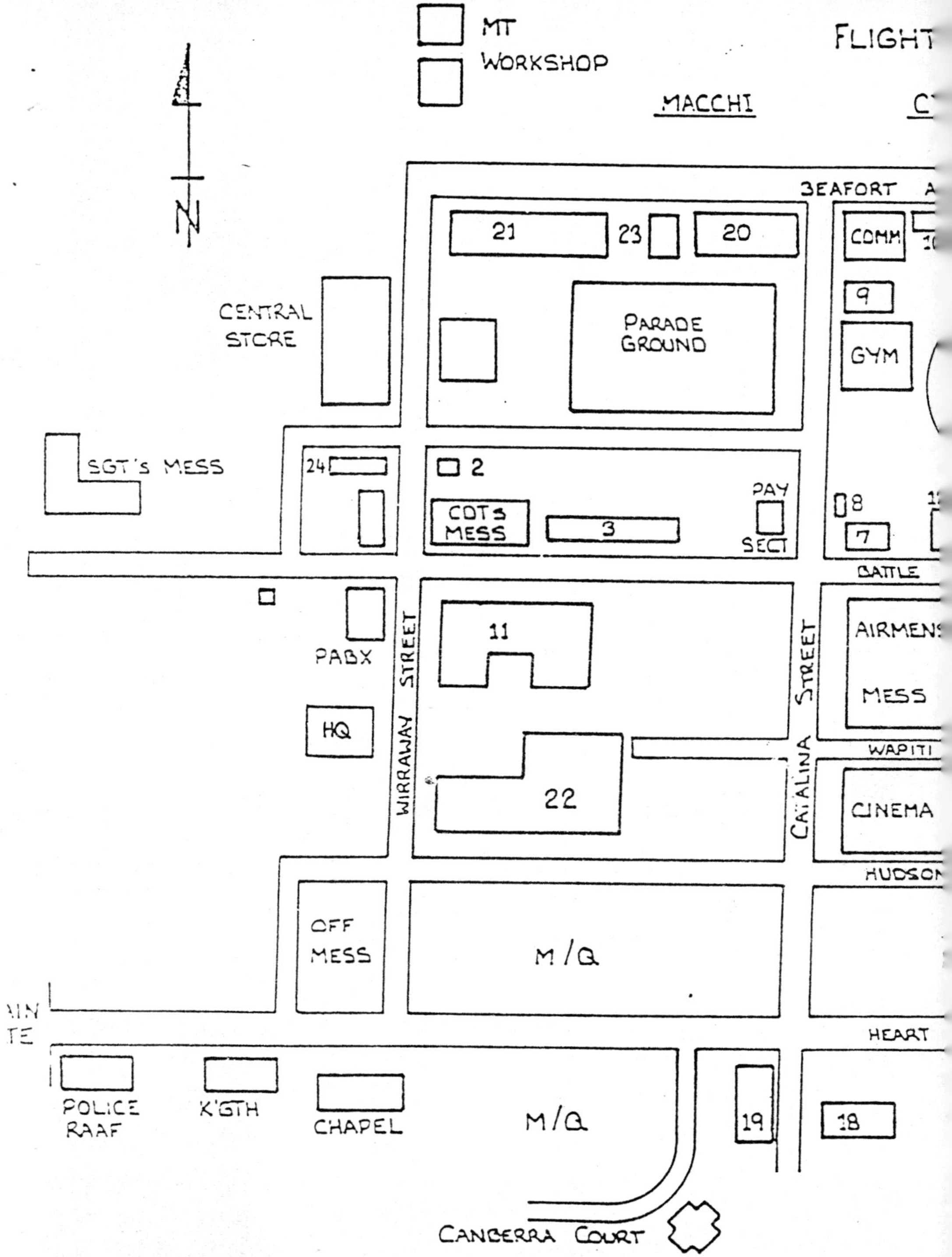
N
MT
WORKSHOP
MACCHI
BEAFORT
21
23
20
COMM
9
CENTRAL
STORE
PARADE
GROUND
GYM
SGT'S MESS
24
2
CDTS
MESS
3
PAY
SECT
8
7
BATTLE
PABX
11
WIRRAWAY STREET
HQ
22
CATALINA STREET
AIRMENS
MESS
WAPITI
CINEMA
OFF
MESS
M/Q
HEART
POLICE
RAAF
K'GTH
CHAPEL
M/Q
19
18
CANBERRA COURT

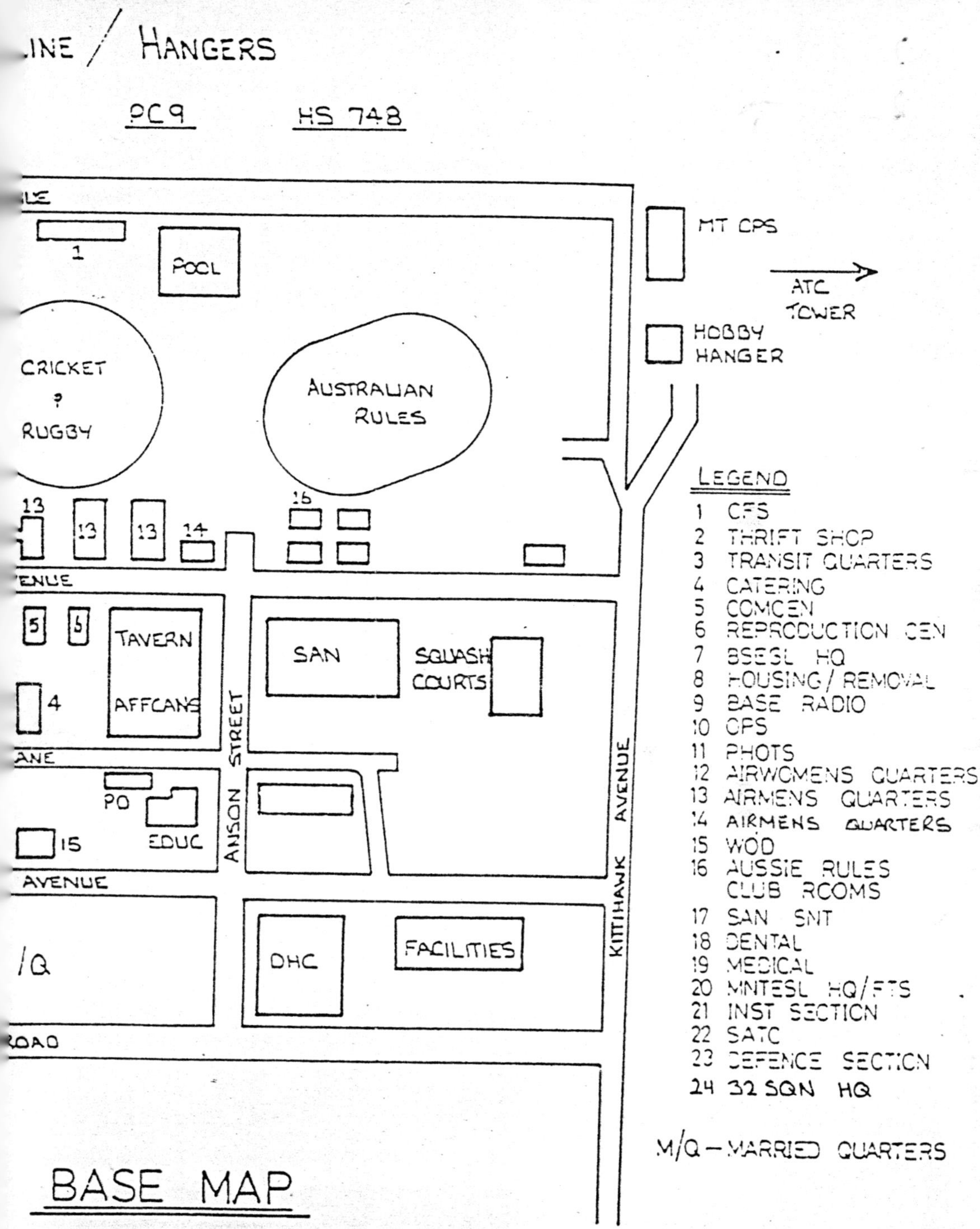
.INE / HANGERS
PC9
HS 748
LE
1
POOL
MT CPS
ATC TOWER
HOBBY HANGER
CRICKET
?
RUGBY
AUSTRALIAN RULES
13
13
13
14
16
ENUE
5
6
TAVERN
AFFCANS
4
SAN
SQUASH COURTS
ANSON STREET
KITTIHAWK AVENUE
ANE
PO
EDUC
15
AVENUE
/Q
DHC
FACILITIES
ROAD
LEGEND
1 CFS
2 THRIFT SHOP
3 TRANSIT QUARTERS
4 CATERING
5 COMCEN
6 REPRODUCTION CEN
7 BSESL HQ
8 HOUSING/ REMOVAL
9 BASE RADIO
10 CPS
11 PHOTS
12 AIRWOMENS QUARTERS
13 AIRMENS QUARTERS
14 AIRMENS QUARTERS
15 WOD
16 AUSSIE RULES CLUB RCOMS
17 SAN SNT
18 DENTAL
19 MEDICAL
20 MNTESL HQ/FTS
21 INST SECTION
22 SATC
23 DEFENCE SECTION
24 32 SQN HQ
M/Q – MARRIED QUARTERS

BASE MAP

The crew of the School's Lincoln 'Gundawarra' at the christening ceremony - 25 August 1947

The School had appreciably developed an armament program since its establishment but staff shortages remained a dilemma for Squadron Leader Slater. Out of a designated manning level of 58 personnel, only 24 staff were on strength which presented difficulties in properly managing the armament ranges at Dutson, Licola and Letts Beach. Vandalistic acts occurred often at the ranges which added to the usual maintenance labour. This low staff level ensued for several years and by 1952, a serious shortfall of single-engine pilots compelled the School to rely upon Central Flying School and their occasional absence effectively cancelled exercises.

In early 1949, RAAF Command authorised Squadron Leader Slater to send specialist instructional staff to No. 82 Heavy Bomber Wing, RAAF Amberley and to the Pearce and Richmond RAAF Stations to give briefs on information amassed by the School and to listen to each unit's suggestions and problems. In June 1956, the Air Armament School introduced a Fighter Combat Instructor Course which taught air combat tactics to fighter pilots who upon graduating returned to their squadrons to impart their knowledge.

On 15 June 1959, Air Armament School disbanded as Headquarters reallocated the School's training commitments to the School of Air Navigation and Central Flying School. Base Squadron at East Sale assumed responsibility for the manning of nearby bombing and gunnery ranges. The School of Air Navigation would conduct the ground training phases for both the bombing section within the Basic Navigator Course and the Recognition Instructors Course, in addition to the ground and air training stages of the Weapons Officer Course. The weapons phase of the Senior Officers Jet Familiarisation Course came under the control of Central Flying School which, together with the School of Air Navigation, took most of the training aids held by the Air Armament School.

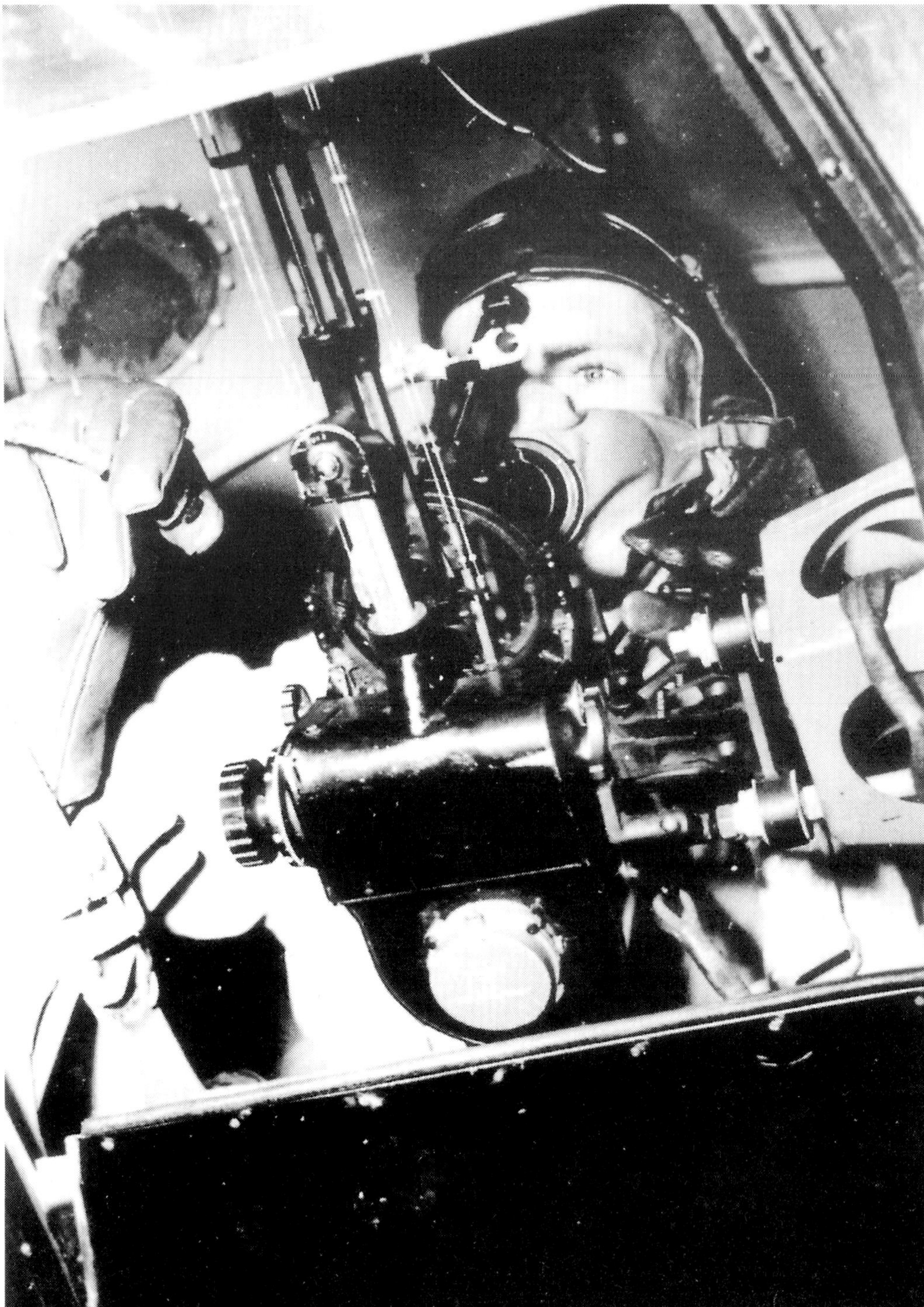

Bomb aimer - such training was done at East Sale

CHAPTER 4

THE SCHOOL OF AIR NAVIGATION

by Flight Lieutenant Greg Haylock

The School of Air Navigation (SAN) was established at East Sale on February 5, 1946. Prior to its formation, Navigators completed courses at an air observers school, a bombing and gunnery school, and an astro-navigation school. Late in 1944, all of these schools were integrated into air navigation schools, but following the disbanding of Bairnsdale's General Reconnaissance School after World War II SAN emerged as the only location for navigation training in the RAAF.

The move from Bairnsdale to the new facilities at East Sale was completed on 14 March 1946. The first Commanding Officer of SAN was Wing Commander J.B. (John) Jewell DFC, who oversaw the development of the School in its earliest years. Wing Commander Jewell had served with No. 10 Squadron RAF since 1939 and had graduated as a qualified Navigation Instructor from No. 15 Course at the RAF's No. 2 School of Air Navigation in October 1941.

The first CO of SAN Wing Commander J.B.Jewell DFC

In SAN's Unit History, he had noted that a considerable amount of obsolescent training equipment had been left in the buildings, and was determined to replace this with the up-to-date equipment required for post-war training. However, even the best training aid at this time, the Celestial Navigation Trainer (CNT), was encountering difficulties getting into full working order, as there were no maintenance operators or instructors left in the service!

The School's combined roles would include: the training of recruits in basic navigation; the provision of navigation refresher courses; the training of navigation instructors, and certain research and development in respect of navigation techniques and equipment.

Visit by 'ARIES'

During 1946, the staff were primarily involved in developing the training section of the school and preparing for the arrival of the School's first course. An interesting and welcome diversion during this time came in November with a visit by the RAF Navigation Mission in the aircraft 'Aries', under the command of Air Commodore N.H. Daith OBE. The members of the mission gave lectures to the School's staff, a number of whom later accompanied the crew of the Aries on a familiarisation flight from Melbourne to Perth and return as an opportunity to handle some of the latest navigation equipment.

'Aries' aircraft at Sale in November 1946

Courses Begin

The first course run at the School of Air Navigation was No. 1 Specialist Navigation Course, which commenced on the 3rd of February 1947 with 11 students. The nine month course was composed of Navigators with extensive war time experience with graduates being employed on the staff of the School, or at Area Headquarters or RAAF Headquarters in Navigation Staff posts. It was noted in the Unit History that this course would pioneer a new idea in lecture rooms with the installation of a 'talkie' projector and an epidiascope in the classroom. This saved undesirable interruption in the instructional procedure caused by moving pupils to some distant projection room.

The course was completed on 7 November 1947, with eight members successfully qualifying as Specialist Navigators. However, due to the relatively small number of navigators required to hold this qualification, subsequent training was carried out at the RAF Central Navigation School at Shawbury, U.K. It is significant to note that three graduates of this one course subsequently became Commanding Officers of the School and/or Officers Commanding the Base.

The first course run at SAN - N 1 Spec Nav, pictured instructing is SqnLdr McCarthy.

Aircraft Arrive

SAN received its first aircraft, Lincoln A73-2, on the 28th of May 1947. It was named 'Nyhuan', which is an East Gippsland aboriginal word meaning 'path finder'. The aircraft was christened on the 25th August, in a ceremony where the electrically-operated champagne bottle breaker failed to operate, and the aircraft captain was forced to scale a platform and break the bottle with a piece of iron. A second aircraft, `GUNDAWARRA', meaning `Spearhead', was also christened as the flagship of the neighbouring Air Armament School. Notably, the electrical fault mentioned was only intermittent; when checked after the ceremony it worked perfectly!

Christening of SAN's first aircraft - 'Nyhuan'

Advanced Navigation Course

On February the 9th, 1948, SAN commenced the first Advanced Navigation Course. The course was available to qualified, proficient Navigators and was designed to provide the RAAF with Squadron Navigation Officers and Navigation Instructors for training units. Thirty three Advanced Navigation Courses were held at SAN, until they were replaced in 1981 with the Aircraft Systems Course and the EW Systems Course. A measure of the prestige enjoyed by the course was indicated by the fact that the Indian Air Force, Royal Pakistan Air Force and the Royal New Zealand Air Force sent officers to undergo this post-graduate training.

Anson Aircraft Arrive

A Refresher Navigation Course commenced on 22 March 1948, with five Anson aircraft being allotted to the School for the flying exercises involved in this training. The course was used to train non-commissioned officers with previous RAAF service who were re-entering the Service. The members were given revision on navigational procedures and an introduction to modern equipment and techniques.

Anson No. 1 over Laverton - similar to those flown at Sale. It was the navigator's job to wind up the undercarriage!

Student Navigators Arrive

The training of Basic Navigation Courses did not begin until the 23rd of August 1948, when 15 students commenced studies on No. 1 Course. Courses ran for 12 months, at the rate of two courses per year. No. 1 Course graduated 11 members on the 26th of August 1949, with a parade that was said to equal similar ceremonies before the war. The parade was reviewed by Air Vice-Marshal F.M. Bladin CBE, and was remembered for its shining bayonets, the gold braid and swords of senior officers, the smart parade work and the music of the RAAF Band.

The parade was also combined with the graduation of No. 1 Flying Training Course. As one former Squadron Leader was reported as saying in the Gippsland Times; 'Times have changed. I remember in the war years when I got my wings, I went to the orderly room one morning and grabbed a pair from a cardboard box. Nothing was said'.

Students from the first three courses graduated from the School of Air Navigation as a Navigator IV, with later courses graduating as Sergeants. Of the 11 graduates of No. 1 Course, nine were posted to Lancaster bombers and two to Dakotas. Several of the course members had served with the RAAF during World War II, including Trainee Navigator R.B. Denniston, who had been awarded the DFC for distinguished services during operations as an Air Gunner. Some of the other members had taken advantage of the RAAF scheme for training selected RAAF ground staff as aircrew.

No. 1 Basic Navigator Course, taken 3 December 1948

Members of No. 2 and No. 3 Navigator courses each attained over 200 flying hours before graduation. The flying training included day and night cross-country navigation exercises, combined with bombing runs. Members of No. 4 course each attained 175 hours, including a long-distance exercise for their final flight. Three Lincoln aircraft flew to Perth for this exercise, while a fourth flew to New Zealand, carrying a trainee navigator from another course to visit his critically ill brother.

A Unit on the Move

SAN has always been a Unit 'going places'. In addition to the flights to New Zealand and Perth, 1948 saw a SAN Lincoln venture as far afield as Japan, transporting the then Chief of Air Staff, Air Marshal Jones. Since then, SAN aircraft have been seen in various parts of the world and, even now, SAN tasks aircraft to Malaysia and each course includes some student training around the South Pacific. Destinations including New Zealand, Vanuatu, the Solomon Islands and Norfolk Island expose students to the challenge of long distance overwater navigation, different Air Traffic Control procedures and also different customs.

Loss of Anson AW-660

SAN lost its first Anson aircraft, AW-660, on the 26th of September 1949. The aircraft was one of three taking part in a familiarisation flight for trainees of No. 3 Basic Navigation Course, but failed to return that afternoon. Three SAN crews searched the coastline north of Seaspray from the air, while members of the Base conducted a ground search in the hills and along the beach, where they found pieces of the wreckage washed up two days later near Lakes Entrance. Five crew members, including two students, were lost in the accident.

Ten years later in October 1959, a jawbone was discovered by a boy on the beach at Seaspray. After pathology tests at the Gippsland Base Hospital determined the bone's approximate age, it was passed to the RAAF Dental Section in Sydney. The bone was identified as that of the Anson aircraft's co-pilot.

Navigator Training

People often visualise navigator training taking place with students sitting in the rear of an aircraft, attempting to maintain a plot of where the aircraft is being flown. However, the RAAF has always used the philosophy that the best way to train a navigator is to have him navigate the aircraft and learn from his mistakes.

Most flying exercises involve one navigator student performing the role of navigator with an instructor providing assistance if required and ensuring that the aircraft is not allowed to stray too far from track. A second student has the opportunity to practice procedures and techniques, whilst not actually controlling the aircraft. The two students usually swap roles half way through the sortie.

Although the instructor is available for advice, students are encouraged to solve their own problems where possible. During the later stages of the course, students may be allowed to 'fly solo', i.e. without an instructor. This will usually help to develop self-confidence as the student accepts full responsibility for the success of the flight.

'Simulators'

To allow suitable training before flights, various ground training devices have been used since 1956. These allow students to practice navigation techniques and procedures on the ground before having to cope with the additional pressures that apply when airborne.

The original navigation trainer was called a dead-reckoning (or D.R.) trainer. Students worked in paired cubicles, with one student conducting a navigation exercise while the second student monitored the actual 'flight path' using an automatic Ground Position Indicator and passed visual, radio and astro fixing information to the navigator. Although of limited capability, the students were allocated 172 hours in the trainer, gaining extra practice without increasing the number of flying hours allocated to the course.

Simulator training at SAN

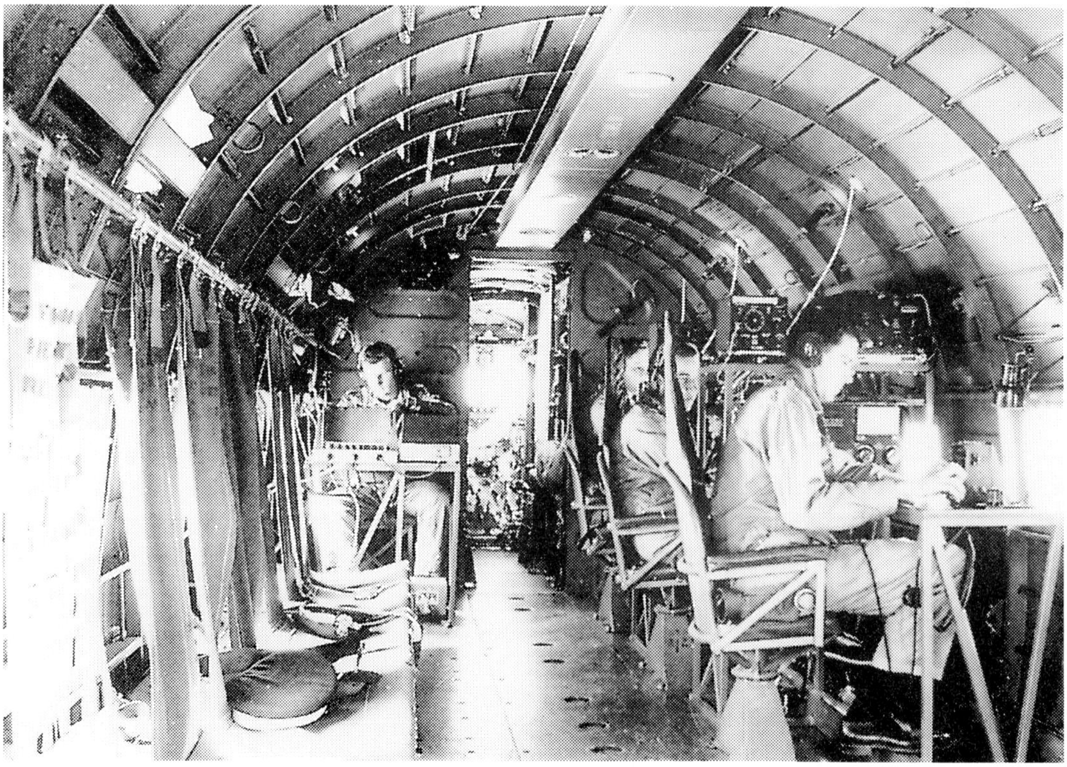

Students at work in the back of a Dakota

A New Aircraft Arrives

Dakotas were first used by SAN in 1950 for No. 5 Navigator course. They took over the Basic Stage flying from the Anson aircraft, which were ferried to Tocumwal for storage in 1951. The relative improvement over the Anson in terms of endurance and serviceability made flying training easier to conduct, but led to problems with an initial lack of pilots. The Dakota required two pilots for each flight, whereas the Anson only required one. Hence, the potential of the six SAN pilots was effectively halved, and required an immediate increase in the Unit's pilot strength.

A SAN Dakota over New Zealand

A Versatile Unit

During its history, SAN has undertaken many tasks far removed from basic navigator student training. In February 1951, Lincoln A73-2 was tasked to deliver urgently needed food and medical supplies to the Antarctic Division Expedition stationed at Macquarie Island. As the island has no airfield, the aircraft was fitted with two long range fuel tanks for the 13 hour, 2800 mile (5 600 km) round flight. This allowed the aircraft more than two hours of flying over the island to drop the supplies. The stores and medical supplies were packed in three `storepedoes', which were fitted to the bomb bays of the aircraft and dropped successfully on separate runs over the target zone.

By complete contrast in climate, on the same day, SAN members manned a Dakota aircraft to patrol all sections of Eastern Victoria spotting bushfires. The aircraft reported several outbreaks directly to the Forestry Commission Patrol Posts in the relevant areas.

In December 1957, a Lincoln with crew members from SAN was used to search for a car believed missing on the Alice Springs - Adelaide road. The crew eventually located the car 11 hours after the aircraft left East Sale, hidden behind trees about 100 yards off the road. The crew directed ground search parties to the vehicle, where they discovered three murder victims in close proximity. News of the discovery made headlines in the national and overseas press.

An Aircraft Type Departs

The Lincoln was phased out from SAN in 1958. Originally a development of the famous war-time Lancaster heavy bomber, the Lincoln gave the School an excellent long-range capability. Group Captain Hodge (Ret), a former Commanding Officer of SAN, recalled an occasion away from Base, when a group of civilians were discussing the Lincoln's appearance. Noting that the Lincoln had two transparent turrets on top, one visitor asked his friend what they would be used for. His friend replied, 'So the navigator can work out where the aircraft is'. Not entirely satisfied, the first visitor replied 'So why are there two?'. Thinking about this, his friend decided 'Well, they carry two navigators on board so one can work out the position of the front of the aircraft while the other works out the position of the rear!'

The final SAN task using Lincolns was to be a four aircraft deployment to New Zealand in April of that year, as part of a single heading and grid navigation exercise with No 14 Advanced Navigation Course. However, on May the 5th a Lincoln aircraft was called out to search for the fishing trawler 'Volga Boatman', missing between Lakes Entrance and Gabo Island. Several crew were involved over three days, but the search was terminated on May the 7th, having found no trace of the trawler.

'Nyhuan' takes to the air in April 1948

Public Relations

Public relations has always featured highly at SAN. Base open days attracted large numbers of visitors, with SAN displaying its early training aids for viewing. The radar training equipment, radio demonstration room and the CNT all drew favourable comments from the impressed visitors. During Air Force Week in 1953, SAN also set up a window display in Grant and Hendricks store in Sale, showing a navigator at work in front of navigation instruments. The proprietors of the store reported that the window 'caused considerable interest amongst the shopping public'.

In November 1962, the newly-formed local TV station GLV-10 spent a day at East Sale, filming proceedings at the School. The footage was broadcast later that month as part of a weekly series called 'This is Gippsland'.

Royal Visit

During February of 1954, all members of the training flight and the instructional staff of SAN were involved in various flying operations associated with the Royal Visit to Australia. Three Lincoln crews carried out escort duties for Her Majesty's Royal Vessel 'SS Gothic', and members of No 8 Advanced Navigation Course took part in Operation 'Royal Salute'. All SAN members took part in the lining of the route for the Royal Party when they arrived by air on March 3rd, and selected members later took part in a display at Point Cook in connection with the visit.

A Move For SAN?

By mid 1955, a lack of manpower and project funding was of major concern. The few navigation instructors were being heavily overworked trying to balance flying, classroom instruction and syllabus development time. SAN had to extensively employ pilots from other units to meet the Lincoln flying program. A proposal to obtain a D.R. trainer had not been actioned, and staff were even proposing to design their own crude model of such a trainer. To add to the problems at the time, sub-zero temperatures were being encountered and several classrooms lacked heating and were affected by inadequate lighting.

Then in October of that year, it was reported that proposed modifications to a classroom had not been approved because of the projected move of SAN to RAAF Base Pearce in Western Australia. No doubt the thought of such a move would have warmed the hearts of staff at the time! The move was probably being considered in order to combine training facilities with the basic pilot course training at Pearce. However, as is well known, SAN did not relocate to a warmer climate. Manning problems were soon rectified and during 1956 many new training facilities became functional. These included the D.R. trainer, LORAN and drift trainers and an H_2S radar trainer.

Air Armament School Closes

Following the closure of the neighbouring Air Armament School, SAN took over the responsibility for weapons training from the 15th of June 1959. This involved conducting the ground training for Recognition Instructor courses and the bombing phase of the Basic Navigator courses, as well as the ground and air training of Weapons Officer courses. Staff and students of the Air Armament School were transferred across to SAN on this date, increasing the size of the school significantly.

Commissioned Officers

Up to 1958, students graduated from SAN with the rank of Sergeant. No. 19 Navigator course, which commenced on the 12th of January 1959, were the first course from which all members graduated as commissioned officers. No. 20 Navigator course saw further changes instituted, as they were the first course to start with cadet status (as opposed to the 'trainee navigator' status of previous courses). In addition, they were the first course to complete all of their training at SAN, whereas previous courses had completed their initial RAAF training at Point Cook.

As a sidenote, No. 20 course's flying training came close to being disrupted in March of 1960, due to the Australia-wide shortage of spark plugs for Dakotas!

Loss of a Landmark

Between 1945 and 1962, the CNT was a well-known landmark on the base. It was described as standing like a medieval fortress on the road between SAN and the School of Photography. The CNT was made by the Link Company of the USA, who made the well-known Link Trainer for instrument flying practice. The primary purpose of the CNT was to train aircrews in navigation by the stars, but it could also be used for bombing training, wireless operating procedures and intercom patter.

Naturally, such a versatile aid was particularly complex. The CNT consisted of a Link Trainer, elevated above the ground, with a 45 foot high wire dome surrounding it. The dome was accurately fitted with small lamps to represent the stars in the night sky over the southern hemisphere. Clouds could even be simulated and sent drifting across the skies. A pilot could actually 'fly' the machine, just like the conventional Link trainer, and even conduct bombing practice in it. For this, a light would flash on a target screen to show where the bomb would have fallen.

The RAAF obtained two other CNTs; one for Rathmines for Catalina flying and one for Tocumwal for heavy bomber crews. However, it is believed that the trainer at East Sale was the only one used successfully, Even so, its use was only intermittent, due to problems obtaining spares and calibrating the delicate machinery.

After about 1955, the CNT featured more as an attraction to small boys on open days than as a training aid. More streamlined techniques were developed in 1957 with the DR Navigation Trainer and as a result, the CNT was hardly used by anyone other than the School of Photography; the tall, angular building being ideal for trainee photographers to practice their perspective shots.

In August of 1962, the CNT was sold at auction, but some discussion remained over whether it had been allocated an aircraft code marking. The 'A' system of code markings was used for all RAAF aircraft, with a third series of numbers starting in 1961. However, A63 was not allocated to any aircraft type, and some academic arguments took place over whether A63 was intended for the CNT before its eventual sale.

No Losses

No. 25 Navigator course, which graduated on August the 15th 1963, achieved a significant milestone. This was the first course to pass out of SAN without any members being suspended. The course even had to endure an all time record May rainfall, which turned areas surrounding the Base into seas, and restricted other flying.

The Celestial Navigation Trainer

Kokoda Trail

In August, 1964, four Dakota aircraft flew members of No. 27 Navigator course to New Guinea for their final stage check flight. During the three day stay in Port Moresby, the students and staff members took part in a familiarisation walk over the start of the Kokoda Trail and surrounding areas. The walk was conducted under the leadership of Major Newman, the Commanding Officer of the New Guinea Volunteer Reserve. In addition, an air exercise was flown across the Owen Stanley ranges, landing at Popendetta and Lae, to familiarise the students and junior staff with flying conditions in New Guinea.

Lunch on the Kokoda Trail for some students of No. 27 Navigator course

SAN Becomes 'Official'

From its conception, the School of Air Navigation used the abbreviation 'ANSCL' in all of its correspondence. A submission was made in August 1965 to adopt 'the traditional and logical abbreviation of 'SAN'. However, this change was not formally approved until 29 July the following year. This coincided with the completion of installing gas heaters to all SAN buildings except for the flight hut. A new toilet building was also constructed around this time; the building is one of the few still standing on the original SAN site.

AEO Training

The RAAF's first Air Electronics Officer (AEO) course began training at SAN in January 1965. The AEO category was introduced to replace the Signaller category, when it became apparent that the term 'signaller' no longer completely described the duties and functions of that role. The traditional task of the signaller - communications - was occupying a relatively small percentage of his airborne time, as operation of the various Anti-Submarine Warfare (ASW) sensors assumed more importance in maritime reconnaissance tasks.

No. 1 AEO course, 2 February 1965

Unlike signallers who graduated as sergeants, the AEOs enlisted as cadet aircrew and graduated as pilot officers in the General Duties branch, with the same career prospects as pilots and navigators. For the early courses, training began at SAN with an emphasis on officer training subjects and an introduction to navigation, meteorology, airmanship and instruments. This covered the first three months of the course, then the students moved to the School of Radio, RAAF Laverton, to begin specialist subject studies. The course lasted for 59 instructional weeks, with one intake of up to 16 cadets each January.

In 1969, SAN took over complete responsibility for AEO training with No. 4 AEO course, using HS748 aircraft. The change from using the Dakota had little effect on the nature of the training, except that AEO cadets now had the opportunity to be trained in radar scope interpretation, which was previously unavailable. As much of the ground training was identical to that of the Navigator course, it allowed the opportunity to program parts of a Navigator and AEO course simultaneously to save on instructor workload. By 1971, the course had been shortened to 47 weeks, making it the same length as the navigator course and allowing the January navigator and AEO courses to start and graduate together.

The introduction of the P3C Orion brought about changes in the nature of the crew's duties and composition. A decision was made to transfer the responsibility for aircraft communications to the navigator, and to create a new position of Airborne Electronics Analyst (AEA) to man the sensor stations. The AEO course was incorporated into Navigator training with No. 56 Navigator course, and AEA training commenced in later years at RAAF Edinburgh. The final AEO course, No. 16, graduated from SAN on June the 28th 1979, in a combined graduation with members of No. 10 Observer course and No. 55 Navigator course. The nostalgic parade was reviewed by Air Vice-Marshal J.C. Jordan, a former Commanding Officer of SAN.

RAN Observer Training

On July 7th, 1969, No. 38 Navigator course arrived at SAN to commence training. This course was unique because it was composed entirely of Royal Australian Navy Midshipmen - the first non-RAAF students to undergo basic training since the School began These students completed the same course as the navigator students before graduating as Observers and joining the Fleet Air Arm of the RAN.

With the introduction of Naval Observer students, Lieutenant Commander Peter Moy RAN was posted to SAN and appointed as the first Senior Naval Officer. Lieutenant Commander Moy was an experienced naval observer, and was the first RAN Observer to undergo the Advanced Navigation Course in 1966.

From No. 39 to No. 43 course, Navigator courses comprised both RAAF and RAN students. Then in September 1972, No. 1 RAN Observer course started training as a separate course. The Observer course involved 37 weeks of ground instruction and 150 hours of practical air training. No. 1 Course graduated on June the 15th, 1973, in a combined parade with the graduates of No. 44 Navigator course. The parade, which had to be held under cover due to weather, was reviewed by the Chief of Naval Personnel, Rear Admiral H.D. Stevenson, CBE.

Dakota Farewell

The Dakota remained in use for navigator training until 1969. No. 36 Course, the last to be trained on the Dakota, graduated on January the 30th, 1969. This was just two weeks after No. 37 Course commenced as the first course to be trained on the new HS748 aircraft.

In September 1973, two Dakota aircraft were handed over to the Indonesian Air Force as part of the Australian-Indonesian program of Defence Co-operation. Two Indonesian navigation instructors received five weeks conversion training at SAN in August/September 1973 before returning home with the aircraft. To mark the end of service for the Dakota a formation flypast took place over East Sale on the 20th of March 1980. The farewell formation was a large Delta shape comprising of three Dakotas, three HS748s, six Macchis and four CT-4 trainers.

No. 1 RAN Observer course with the CO, Wing Commander Reg Gillard and SNO, Lieutenant Commander Tapping - 28 September 1972

Dakota farewell flypast 20 March 1980

HS748 Navigation Trainer Arrives

Throughout 1967, preparation was under way for the changeover of aircraft from the Dakota to the Hawker Siddeley HS748. Eight of these aircraft were ordered as replacements for the Dakotas at SAN, with the insides having been specially designed for navigator and AEO training.

The first crew departed Australia for London in June 1968 to be converted onto the aircraft and to ferry the aircraft back to Australia. Wing Commander Staib, the Commanding Officer of SAN, was the leader of this detachment. He returned aboard the first aircraft, A10-601, which flew from London via Athens, (where the crew were delayed for five days due to lack of diplomatic clearances), New Delhi, Bangkok, Singapore, Indonesia, Darwin and Alice Springs.

A10-601 arrived in Sale on September the 20th, 1968. There was no official ceremony to mark the occasion, but the aircraft was met by the Officer Commanding, Group Captain A.R. Hodge, as well as the Mayor of Sale, the President of Avon Shire and other local dignitaries. The reception of the aircraft was conducted in a rainstorm, before the guests and crew retired to the Officers Mess for afternoon tea.

A10-602 arrived on October the 12th, 1968, with A10-603 arriving in November. Student training on the HS748 began with No. 37 Navigator course, which commenced at SAN in January 1969. Before their flying training commenced, senior staff members took one HS748 on a route proving flight to New Zealand, returning via Norfolk Island. This route would remain a standard long-distance overwater test for many years.

A High-Powered Welcome

Wing Commander Staib recalled an incident that occurred on the next visit to Norfolk Island in October 1969. This was during the return of No. 23 Advanced Navigator course from New Zealand, the first Advanced course to train in the HS748. After the first aircraft arrived on the island, Wing Commander Staib was keen to show the Island Administrator over the new aircraft, whilst the captain of the aircraft waited to marshal the second aircraft into the confined parking space. As the CO was proudly explaining the navigation set-up, he was called to the door of the aircraft. The embarrassed aircraft captain reported that the tail fin of the second aircraft had broken an overhead electrical cable, and asked the CO to pass this information on to the Administrator. The crew found out later, to their further embarrassment, that this power cable served the northern part of the island, which was left without power for several hours!

Norfolk Island Connections

Despite this incident, SAN has continued to be welcome at Norfolk Island and has maintained some close contacts with the people there. These began during the Dakota era, when on one occasion there was no motel accommodation available on the island and the crews were home billeted with local families. Squadron Leader Wyatt, a former OIC SAN Flying Flight, recalled that all members were extremely well looked after and that Norfolk Island was regarded by SAN as a 'home away from home'.

In another connection with Norfolk Island, the Administrator in the earlier story was Air Commodore R.W. Dalkin, who had been the Chief Instructor at the General Reconnaissance School, Bairnsdale when Wing Commander Staib was undertaking the course in 1944 The position of Norfolk Island Administrator was later taken over by a former Commanding Officer of SAN, retired Air Commodore E.T. Pickerd OBE, DFC. As a Wing Commander, he had commanded the School from 1955 to 1957 and continued on a distinguished RAAF career before taking up the appointment as Administrator in 1972.

Dakotas at Norfolk Island

Another Norfolk Incident

Several years later in 1985, a Norfolk Pine tree inadvertently jumped into the way of A10-601, which was manoeuvring into a parking position. As the tree only caused superficial damage to the aircraft's wingtip, the aircraft was able to be flown back to Sale after a thorough inspection. The aircraft's captain, who was later posted to fly fighter aircraft (with smaller wingspans), tried to claim that he had in fact improved the aircraft's performance as it was subsequently found to fly 10 knots faster than the rest of the HS748 fleet!

Synthetic Navigation Trainer

The eighth and final HS748 aircraft arrived from the U.K. on August the 7th, 1969. To aid with preparing students for flights, a Synthetic Navigation Trainer (SNT) was ordered. The SNT comprised 12 cubicles, each set up to reflect the Nav 1 station in the HS748. Each cubicle operates independently, being computer driven and capable of simulating the navigation information that a student would receive inflight. The SNT project ran a year overtime, with the official handover from Hawker Siddeley Dynamics Ltd finally taking place on March the 17th, 1970.

Cadets Mess Opens

In addition to having new aircraft, another highlight of 1969 was the opening of the new Cadets Mess. Students occupying the Mess in July, but the official opening took place on August the 14th, 1969. The Air Officer Commanding Support Command formally opened the Mess, with a small function being held afterwards to mark the occasion. On August 30th, a formal dance was held in the Mess, which was the first occasion for over a year that there were sufficient Mess members to run such a function. The event was a great success, and only the stronger spirits managed to attend an impromptu barbecue lunch the next day. However, the Cadets Mess did not become fully functional until May 18, 1970 when all of the mess facilities became available.

Opening of the Cadets Mess 14 August 1969

One feature that caught the eyes of visitors to the Cadets Mess was the pair of Japanese swords that were mounted on the panelled wall. The swords, one a Samurai sword and the other a naval sword stamped 1895, were presented to the Mess in 1975 by Group Captain Staib, then Officer Commanding RAAF East Sale. Group Captain Staib had obtained the swords 30 years earlier in Awakuni, Japan, as a member of the British Commonwealth Occupation Forces.

Another prominent feature was the massive hedge that surrounds the Cadets Mess on three sides. Its imposing nature prompted many infamous 'hedge races', where members would climb the hedge and race from one end to the other. Competitors were often seen to disappear completely within the hedge, while the braver and hardier ones would even attempt to jump the gap in the hedge where the footpath crossed through. Despite the nature of the event, very few major injuries were recorded.

A Surprising Student

One student from No. 39 Navigator course surprised SAN staff when it became known that he had been secretly married for three months! Headquarters Support Command approved his retention on course, and the student later not only graduated, but was awarded Dux of the Course and received three other trophies for his performance.

Eight Aircraft Flypast

A formation flypast of all eight SAN HS748s took place on June the 21st, 1971. SAN captains had been practising formation flying for the graduation parade of No. 40 Navigator course, then made the historic flypast four days later when Maintenance Squadron were able to provide eight serviceable HS748s, and one Dakota for photography. Unfortunately, the photographs taken from the Dakota did not turn out particularly well. The exhaust haze coming from the engines interfered with the shots, leaving the privately-taken photos from the ground as the best record of the event. As a sidenote, the co-pilot of aircraft A10-606 for the flypast was Pilot Officer P.J. Cleary, who would later return to Sale as the Commanding Officer of No. 32 Squadron when that unit reformed in 1989 equipped with all ten RAAF HS748s.

Cyclone Tracy Relief Support

Most Australians are aware of the role that the RAAF played in providing relief support following the tragedy of Cyclone Tracy in December 1974, but few people are aware that SAN sent two HS748s to assist with the transporting of families out of Darwin. The crews departed on December 28, returning on December 31 and January 1. One HS748 also flew a half ton of food and baby clothing collected in Traralgon, to Laverton where it was transferred to a Hercules aircraft bound for Darwin.

High Performance

As well as having a sound maintenance record, the HS748 was also shown to have surprising flying abilities. One Friday afternoon, A10-607 performed a 'barrel-roll' manoeuvre, something not normally seen in an aircraft of this type. The manoeuvre was completed within the airframe structural limits, and the captain noted it did not even spill water out of the toilet.

In 1984, a HS748 was invited to the World Aerobatics Championships held at Ballarat - but the aircraft took part as a static display only and did not participate in the competition.

Low Performance

On April 13, 1982, No. 16 Observer course began training at SAN with two members. This course would later be remembered as the first to produce no graduates. One of the members, Midshipman Tony Leggett, was suspended at his own request and later left the Navy and joined the Bureau of Meteorology. He returned to East Sale several years later to work in the weather office, and, as part of his work with the Bureau, instructed several Navigator and Observer courses in Meteorology.

Russian Rescue

In August 1977, a crew from SAN was involved in an unusual rescue off the West Australian coast. The crew was at RAAF Base Pearce on the last leg of a staff navigational trainer, when approached to assist the Russian trawler 'Filoent', which had a critically ill woman on board. The crew's task was to firstly locate the trawler, which was about 50 km off course, then support the rescue Iroquois helicopter because of its limited navigation facilities.

The operation, carried out in 30 km/h winds and a 4.5 to 6 metre swell, took over three hours, including an hour to locate the trawler - a difficult task because of the high seas and poor visibility. An important member of the rescue crew was Wing Commander Kevin Egan of Pearce who spoke Russian and instructed the non-English speaking 'Filoent' crew to manoeuvre the trawler into position for the rescue. Even so, the helicopter's task was risky and when the doctor was lowered to the ship by winch, the ship rose in the swell causing the doctor to hit the deck heavily. The SAN crew members were later advised that this rescue could not have been successfully completed without their assistance.

SAN Receives Queen's Colour

SAN received the Queen's Colour, a light blue silk flag with the School's Unit badge in the centre, on September the 28th, 1978. It was presented by the Governor-General, Sir Zelman Cowan, in a dual presentation with Central Flying School. The dual presentation was believed to be unique in Australian Service history. The following day, both Units exercised their right to march through the City of Sale, combined with a flypast of aircraft from SAN and Central Flying School.

The Russian trawler `Filoent'

Presentation of the Queen's Colour to SAN and CFS

AC Trouser

For several months in 1978 and 1979, SAN adopted an unusual mascot. Aircraftsman (AC) Trouser was a two metre long carpet snake of unknown sex. It was known to often entertain crews at SAN, and to occasionally petrify guests. Its foster parent was Flying Officer Brian 'Speedie' Meade, a captain on the HS748, who said the snake's name was chosen after 'considerable deliberation'. AC Trouser showed quite an interest in aircraft, even managing to get stuck inside a heater that was switched on, before reluctantly being removed three hours later.

SAN's mascot - FlgOff 'Speedie' Meade with AC Trouser.

Bicentennial Air Race

As part of Australia's Bicentennial celebrations, an HS748 from SAN took part in the GE Bicentennial Around Australia Air Race in September and October of 1988. The race started in Narromine, NSW, and travelled north to Darwin, through Broome to Perth, then back through Adelaide and Melbourne before finishing at a civic reception in Canberra. The race involved accurate navigation at low level, with all aircraft being given a handicap according to size and speed. The SAN crew, led by the CO of SAN Wing Commander L.D. Noble, performed extremely well to finish second overall, and were the highest placed military crew. Much of the credit for the success belonged to the maintenance crew, who worked many long hours each night to keep the aircraft serviceable.

SAN crew for the Bicentennial Around Australia Air Race

Post Graduate Training

As well as the training of basic courses, post-graduate training has always held a major role at SAN. The first two courses held at SAN were post graduate courses, being the one-only Specialist Navigation Course and the longer-running Advanced Navigation Course. Up to 1980, several other types of courses had been conducted at SAN to meet the needs of the RAAF and even civilian agencies.

In August 1956, No. 1 Fighter Squadron Navigation Officers Course commenced training with six junior pilots. The course ran for six weeks, training the fighter pilots in navigation techniques applicable for their role. No. 2 Course was held in September and October of 1960.

The School has conducted several Antarctic Navigation Training courses for members of the RAAF Antarctic Flight. The course was designed to prepare graduates for operations with the Australian National Antarctic Research Expedition. Six courses were held between September 1956 and October 1960, being of three to four weeks duration. Many years later in 1976, SAN also provided polar air navigation training for three civilian members of the 1976/77 Australian Antarctic Expedition, who were flying light aircraft and helicopters in support of the expedition's scientific program.

SAN has conducted many other courses, some existing longer than others. In May 1977, SAN commenced the first of five Navigator Communications courses, to train navigators in airborne communications prior to undertaking the P3C Orion Nav/Com conversion course. In January 1978, No. 1 Department of Transport (DOT) Navigation Course was held to give DOT instructors an appreciation of RAAF Search-and-Rescue methods, procedures and capabilities. Astro Navigation Conversion courses were run from 1951 and 10 Senior Weapons Officer courses were also conducted during the 1950s. In later years, SAN conducted three Pilot Navigation courses, commencing in 1976, to graduate pilots for appointments as Unit and Staff Navigation Officers.

As described earlier in this chapter, the Advanced Navigation Course was the second course to commence at SAN, in February 1948. Initially designed for qualified navigators, the course was later changed to make it available to all aircrew officers. The changing needs of the RAAF saw the Advanced Navigation Course being supplemented by a Weapons Systems Course in 1977, and an Electronic Warfare Systems Course in 1982. To more accurately reflect its changing nature, the Advanced Navigation Course was renamed the Aircraft Systems Course in 1982. In 1986, responsibility for conducting the Electronic Warfare Systems Course was transferred from SAN to the newly formed Electronic Warfare Operational Support Unit in Adelaide.

No. 32 Squadron Re-forms

1989 was a year of many changes at SAN, but nothing could have been bigger than the reforming of No. 32 Squadron at East Sale on July the 1st. On this date, all pilots of the SAN Flying Flight were transferred to No. 32 Squadron, along with the HS748 aircraft. While navigator training remains the major role, the formation of No. 32 Squadron enabled the aircraft to be more effectively utilised for general transportation tasks.

To allow No. 32 Squadron to balance these two roles, SAN adopted a policy of using a maximum of three aircraft per day. At the same time, SAN changed its method of training from running two Basic courses per year to running four smaller courses (with a maximum of 12 students) per year. This reduced the number of days each course would require to complete a particular flight using only three aircraft. Gone were the days of the six aircraft 'stream' navex, so common in the past.

The 'Hoppy' Roos Trophy

On June 16 of 1989, just before the formation of No. 32 Squadron, tragedy struck at SAN. One of SANs most popular pilots, Flying Officer Bart 'Hoppy' Roos, died as a result of a car accident while travelling home on his bicycle after returning from a flight. In his memory, a sporting competition is now held each year between SAN and No. 32 Squadron, with the victor taking possession of 'The 'Hoppy' Roos Memorial Trophy'.

New SAN Building Opens

When training recommenced in January 1990 after the Christmas break, staff and students were able to enjoy first-class facilities in the School's new building. The two-storey design meant that all of the School's members were now under the same roof, instead of being spread amongst many temporary buildings. The new building also accommodated the SNT, a secure area for weapons and systems course training, and two astro training rooms at the top of the building. These two rooms allow students to practise using the sextant whilst on the ground through the aircraft-style sextant mounts in the roof. Previously, the sextant mounts were attached to metal frames set up in the courtyard of the old SAN site, which meant some extremely cold nights for those students practising their night astro techniques in the middle of winter!

CO, Wing Commander 'Iggy' Noble and Air Marshall Funnell at the opening of SAN in May 1990

The new building was officially opened on Friday May the 18th, 1990, by the Chief of Air Staff, Air Marshal R.G. Funnell, AC A reunion of SAN staff and students was also held over the period 18-20 May, with former members being flown in from as far away as Brisbane. A members dining-in was held in the Officers Mess on the Friday night and several mixed functions took place in the new SAN building during the weekend. The whole occasion was well attended and included guests from the early days of the Specialist Navigation Course.

Former COs of SAN at the Reunion Dining-in-Night, May 1990.

(L-R) E.T Pickerd (55-57), K.M. Staib (67-69), G.L. Barlow (85-87), D.R. Marks (73), K.R. Blakers (78-79), H.A.H. Pickering (55), L.D. Noble (87-90), R.E. Gillard (71-73), A.R. Hodge (50-52), P.W. Kennedy (74-76)

Combined Navigator/Observer Training

Navigator and Observer courses regularly started together, as the initial training was the same for both courses. After about the fifth flight, the two courses would separate and concentrate on their own requirements. Gradually the two courses became more common to each other, until 1990 where the decision was made to combine No. 78 Navigator and No. 27 Observer Courses entirely. This meant that for the first time, Naval Observer students were taught astro navigation and RAAF Navigator students were taught maritime overwater procedures - this 'cross-training' being beneficial to all students.

Female Students

On June 4, 1990, another first happened for SAN. Lieutenant A.J. (Mandy) Goodier, RAN, commenced training on No. 28 Observer course, becoming the first female student to undertake basic training at SAN. She graduated in June 1991 and was posted to HC 723 Squadron at HMAS Albatross in Nowra. Shortly afterwards, Officer Cadet Jocelyn Harvey became the RAAF's first female Navigator student when she commenced training on No. 84 Navigator course in July 1991. Jocelyn graduated on June the 4th 1992 and was posted to No. 37 Squadron in Richmond.

The only major problem that these two students found during their training was that the stands used when operating the sextants were designed for the average male student height. Being slightly shorter than this, both resorted on occasions to standing on either a couple of thick books or even the sextant box in order to reach the sextant. These two students were not the first female aircrew members at SAN. Flight Lieutenant Robyn Williams, one of the first female pilots to graduate in the RAAF, had already been posted to SAN after completing pilots course in 1988.

SAN's First RAAF Female Graduate - AVM Gary Beck congratulates OffCdt Jocelyn Harvey

A Familiar Face Retires

July 17, 1990, saw the staff farewell a member who had served SAN almost as long as the HS748. Mal Adamson had been the civilian general hand at SAN since 1972 and saw many students pass through SAN and return in later years as instructors.

As well as performing cleaning duties and building maintenance, Mal looked after the gardens and lawns around the School and took particular care of the 40 or so rose bushes. In 1973, the School put several entries into the rose section of the Sale District Agricultural Show, winning two second places and the Champion Rose of the Show. The roses were originally planted by Flight Lieutenant Frank Hill, an Instructor at the Air Armament School. The Officer Commanding East Sale at that time, Group Captain A.R. Hodge, recalled that Frank Hill travelled far and wide around Gippsland, visiting almost every nursery in the area in the elusive search for a blue rose to complete his collection. In 1990 when SAN moved into its new building, Mal made sure that 11 of the rose bushes, in a variety of colours, were moved into a bed at the entrance to the building just behind the sundial. These rose bushes now act as a reminder of his lengthy service at SAN.

Inertial Navigation Systems Arrive

In keeping with the increasing technology of the RAAF's operational aircraft, the HS748 fleet undertook an upgrade to its navigational equipment with the installation of the Litton LTN92 Inertial Navigation System (INS) during 1991/92. The upgrade took approximately six weeks per aircraft and was required largely due to the difficulty of maintaining the ageing Ground Position Indicator GPI Mk7. The increased capability meant changes to training procedures as some careful planning was required in determining when the students of No. 83 Navigator, No. 30 Observer course and No. 84 Navigator course would change from using GPI-equipped to INS-equipped aircraft. The update program also involved removing the communications station from the rear of the aircraft, which had not been required since the deletion of dedicated communications training from the syllabus.

To continue the routine of providing procedural training for the students before each flight, the SNT also required a major upgrade to enable it to run the Litton 92 INS. The contract was awarded in July 1991 with a completion date set at Feb 1992, but like the original SNT contract the project ran several months overdue.

A New Era of Training

During 1990, a series of reviews culminated in SAN being tasked to produce a new Basic Navigation Course which philosophically was very different from the previous one, having a greater emphasis on crewmanship and air awareness. The resultant course is six to eight weeks shorter than the old course and has the assessment emphasis on application rather than knowledge of procedures. The course is also split into two phases; the basic phase and an advanced 'role-reinforcement' phase. The basic phase consists of 16 flights covering the majority of the training of the former syllabus. On completion of the basic phase, students are then advised of their initial end-of-course posting locations and divided accordingly into either the Tactical or the Maritime role-reinforcement phase. Naval Observers automatically follow the Maritime course, while the RAAF students' postings are determined by their performance and aptitudes during the basic phase, their own preference and the availability of positions within squadrons.

The 10 week advanced phase covers navigation techniques specifically related to the Maritime or Tactical role. The Maritime role involves Search and Rescue, Maritime surveillance and area operations. The Tactical role uses the PC-9 trainer, flown by instructors from the Central Flying School, for low-level visual navigation exercises and the HS748 for medium-level radar navigation exercises. The two groups are then brought back together for their graduation parade. This new course came into effect with No. 85 Navigator/No. 31 Observer course, which commenced in September 1991.

Further Changes to SAN Training

During the same time frame, SAN undertook a major review of all its post graduate training with a view to identifying and better meeting the customer's requirements. As a result of this review, in 1992 SAN commenced three new post graduate courses; the Weapons Employment Course, the Airborne Instructional Technique Course and a remodelled Navigation Instructor Conversion Course. The five week Weapons Employment Course provides aircrew, engineers and intelligence officers with training in weapons guidance, weapon effects and weapon effort planning. The three week Airborne Instructional Technique Course provides a method for training any aircrew members who will be involved in instructional duties, either within SAN or on Squadron conversion courses. In 1993 a further new course, the Aircraft Integrated Systems Course, also commenced. This course will include a distance education component, another new venture for SAN.

As well as the revamp and replacement of SANs traditional courses, SAN has taken on new responsibilities. These include the Airman Aircrew Basic Course, Basic Intelligence Officer Course and Intelligence Analyst Course. At the time of writing, these courses are all under development and will commence at SAN in 1993.

SANs international profile has also increased over the 1991/92 period and shows promise of continuing to do so. HQADF has tasked SAN with a twice yearly lecture commitment in Malaysia in support of the Five Power Defence Arrangement. This involves a course comprising of officers from Australia, United Kingdom, New Zealand, Malaysia and Singapore. Additionally, 1993 will see the incorporation of Royal New Zealand Air Force (RNZAF) Basic Navigator training and Airborne Electronic Operator (AEOp) training into SAN. The RNZAF will also be providing instructors to augment SAN's current staff.

The net result of the changes over the last eighteen months, and those to come during 1993, is a larger, higher profile School with a 40% increase in student throughput and an international reputation as a centre of excellence in navigation, weapons and aerospace systems.

After 47 Years

The following sentences were written by Air Marshal R.G. Funnell, AC, in his foreword to the booklet accompanying the official opening of the new SAN building in 1990:

'SAN has maintained its place in the force structure because it has been able not only to adapt to the changing world of military aviation but also to so position itself conceptually as to direct change in beneficial directions. Consequently, although the equipment and the operating techniques and procedures of navigation have changed profoundly in the 44 years of the School's existence, SAN continues to fulfil its charter and conduct its valuable work'.

Throughout these changes, the School's badge and motto have remained appropriate for all of its students. The motto, 'Wisdom Through Diligence', reminds us that wisdom is gained through knowledge, a product of consistent and applied hard work. The badge of the School approved by King George VI and registered in 1948, contains a torch blazing, indicative of learning, on a background of the Southern Cross. As well as being symbolic of Australia, the stars remain a symbol of navigation and indicate the importance that astro navigation has held and continues to hold at the School.

Perhaps as a result of increasing technology, in the future the stars on the badge may relate to SAN in terms of navigational satellites orbiting the earth; but that would once again show SAN adapting to the changing world of military aviation. But no doubt the badge will remain the same, reflecting the philosophy stated by Air Marshal Funnell: 'Much that is of benefit today and will be of benefit tomorrow can be gained from knowing where we were and what we were doing yesterday'.

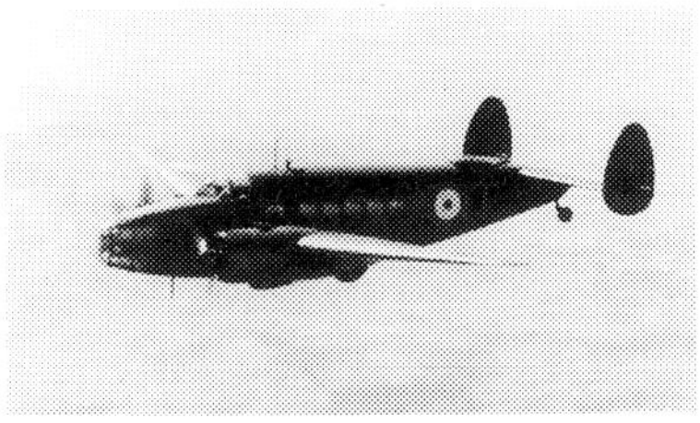

CHAPTER 5

CREW CONVERSION UNIT

by Flight Lieutenant Andrew Layton

The new year of 1946 ushered in a new Commanding Officer and two infant units to RAAF Station East Sale. In February, a nucleus party of the disestablished Air Gunnery School arrived from RAAF Station West Sale to found the Air Armament School and the incoming Commanding Officer of East Sale, Group Captain Deryck Kingwell, welcomed that detachment and a vanguard contingent of the Crew Conversion Unit (Heavy Bomber).

The Crew Conversion Unit had emerged from the remnants of No. 7 Operational Training Unit and cleared out of RAAF Station Tocumwal to instruct heavy bomber crews at East Sale. This development was part of a RAAF wide reorganisation of training for the post-war Air Force. The waning requirement for training had necessitated an integration of existing training schools and their subsequent merger into a minimum number of training units.

On 15 February, a Liberator landed at East Sale from Tocumwal conveying some of the Unit's personnel and they immediately began to prepare facilities for the maintenance wing and offices for the headquarters section. The Commanding Officer of the Crew Conversion Unit, Wing Commander William Brill, came to East Sale three days later to inspect the complex and near the end of February, several aircraft ferried the remaining unit personnel and stores to East Gippsland.

The urgency with which the Unit embarked to East Sale was perhaps unnecessary as the staff waited until October to collect a Lincoln from No. 1 Aircraft Depot, RAAF Laverton. The acquisition of Lincoln A73-5 though was no insurance that the Unit would increase its workload as there were insufficient maintenance personnel to service the aeroplane. Because of this problem, Wing Commander Brill sent a small ground crew to Tocumwal to arrange a maintenance service for the second Lincoln assigned to the Unit.

In December, a tasking role was given to the Crew Conversion Unit when the Air Armament School requested 25 Liberator flying hours for dual circuits and landings toward fighter affiliation periods on its Pilot Gunnery Instructors Course. The Unit would also fly regular bushfire patrols over East Gippsland during summer as No. 1 Operational Training Unit had discontinued these sorties upon its disbandment. To support the Victorian Forestry Commission, a Lincoln from the Unit flew nine bushfire patrols and reported 44 fires during February 1947.

The instructional staff were also engaged in understanding and installing the Lincoln training equipment. Flight Lieutenant Rex Whitburn started a Lincoln conversion with Group Captain Kingwell later joining him on course. Both pilots completed their night conversion in the left and right hand seats by April.

Crew Conversion Unit Staff - 15 Jul 47
CO Wing Commander Bill Brill is second from the left,
Squadron Leader 'Rusty' Steele is on his left.

On the 5th of May, two crews arrived from No. 82 Wing, RAAF Amberley, on attachment for a Lincoln Conversion Course but this training commitment only partly compensated for the indecisive administrative direction given to the Crew Conversion Unit. There were no airmen posted to the unit which had forced the Chief Ground Instructor to temporarily abandon his primary duties and attempt to install some of the training apparatus. In the unit history report, Wing Commander Brill wrote of the Unit's predicament: 'Although this unit is to be the Heavy Bomber Training Unit of the RAAF, we still lack a policy on which type of aircraft will be used and moreover, what types of equipment will be installed in the aircraft to be used. This makes the setting up of any training aids doubly difficult.'

The unit's unknown future became clear in July when the RAAF authorised its disbandment by the first day of August. Before then, No. 3 Lincoln Conversion Course graduated at the end of July and staff were gradually posted out of the unit. On the 27th of August 1947, the Crew Conversion Unit (Heavy Bomber) ceased to exist in the post-war Air Force, formally heralding the end of the era of big bombers.

A view of Lincoln A73-8's bomb bay laden with 6,000 lbs of bombs. On 22 Jan 48, Flight Lieutenant F. Montgomery flew this configuration with what was regarded as the heaviest bomb load ever carried by an aircraft in Australia

CHAPTER 6

HEADQUARTERS RAAF EAST SALE

by Flight Lieutenant Andrew Layton

The cessation of hostilities signified major changes for the personnel and equipment at East Sale. On 8 September, a census was taken to determine the wishes of all students and those who elected discharge immediately withdrew from their course. Seventy-seven students chose to leave the RAAF and were posted to personnel depots for discharge within a fortnight. This initiative did not upset the balance of the courses and No. 33 Hudson General Reconnaissance Course graduated three crews on 25 September. The crews departed East Sale on posting to No. 1 Aircraft Depot, Laverton and it was the last Hudson Course held at East Sale. The students of No 30 Beaufort Transport Course and No. 30 Torpedo Attack Conversion Course were not given an opportunity to graduate as the RAAF terminated their courses two days later and training stopped at No. 1 OTU.

RAAF Station East Sale in 1943

Demobilisation

The end of flying training on the station brought a reorganisation to the Maintenance Wing. The wing was producing more serviceable aircraft than required for the flying program but an increase to the number of aircraft stored in the serviceable pool had necessitated more personnel to service and maintain the aeroplanes. Maintenance man hours on each aeroplane had not appreciably eased as engines had to be turned daily, started and run weekly with covers for engines, airscrews, cockpit and gun turrets removed. If aircraft were not moved regularly the tyres developed flat spots or went out of shape and the maintenance technicians regularly checked for corrosion in the aircraft and updated records and log books. All the Beauforts were transferred from the dispersal area and parked in the hanger space alongside Hudsons, Venturas, Vultee Vengeances and Kittyhawks for servicing and repairs.

For those personnel at East Sale not under training, life would involve more leisure time as No. 1 OTU had a new work schedule that included two standown days per week. Dances were still staged in the station gymnasium every Thursday to the strains of the unit orchestra but less people attended as postings out of East Sale increased. Sports fixtures continued but were being scaled back as the stable of sporting champions who once lived at East Sale cleared out of the unit and left the RAAF. Publicised as probably the last boxing tournament between the Air Gunnery School, General Reconnaissance School and No. 1 OTU, the event featured a novel blind folded bout and a wrestling contest between Sydney's Warrant Officer Woodhill former Heavyweight Champion of Australia, and Warrant Officer Williams of No. 1 OTU who was awarded the match on points.

A few days after this event, a discharge section opened in the former Training Wing Headquarters to reduce the congestion at Unit Headquarters caused by personnel leaving on posting to personnel depots. The section's staff handled clearances and a rehabilitation officer helped members to complete the myriad of rehabilitation forms. As a farewell gesture to personnel stationed at East Sale for the past few years, No. 1 OTU staged a Victory Ball in the gymnasium on 2nd November. A floor show accompanied by twin orchestras entertained the guests who enjoyed one of their last functions as RAAF members.

The thoughts of discharge briefly left the minds of East Sale personnel when His Royal Highness, the Duke of Gloucester and Governor General of Australia, arrived at East Sale to congratulate No. 1 OTU for its commitment to the Second World War. The unit assembled a guard of honour on the tarmac and the Commanding Officer, Group Captain Keith Parsons, the Mayor of Sale and the President of the Avon Shire Council, greeted the Duke upon his arrival. The Duke had visited Sale in 1934 to plant a Royal Oak Tree in Victoria Park and later toured the Sale Agricultural Show after the inspection of the station.

At this point in the history of the RAAF, there was a reduced demand for training of personnel in the post-war Air Force. Within a year of the Second World War ending, the Air Member for Personnel, Air Commodore Joe Hewitt, administered the disengagement of 140,000 service men and women from the RAAF and WAAAF which left a mere 7,000 personnel in uniform.

This period of East Sale's 50-year history provided happy memories for many people but the rapid rate of discharges was adversely affecting the station's messing. Many airmen believed that discharges had decreased the number of messing officers which led to a decline in the quality and quantity of meals served in the Airmens' Mess. These disgruntled airmen endured the meals for a few weeks until they staged a 24-hour hunger strike on 5 December. One airman said in the Gippsland Times, "if the cooks lose the tin openers, we will starve." Group Captain Parsons ordered a parade afterwards to hear the grievances and although the airmen thought some enrichment in the food had resulted from the strike, further improvement did follow.

Demobilisation of service members was under way by November and onwards to December, 358 personnel left East Sale on posting to personnel depots throughout Australia. The end of the war had also rendered several military aircraft redundant and 58 Beauforts were ferried to No. 5 Aircraft Depot, RAAF Station Wagga and another 10 Beauforts flew to No. 7 Aircraft Depot, Tocumwal. These aircraft were put in storage in addition to seven Hudsons, five Vultee Vengeances, four Kittyhawks and 10 Oxfords which were delivered to aircraft depots at Richmond, Laverton, Oakey, and Western Junction respectively.

In December, the Air Board initiated a reorganisation and amalgamation of existing training units to produce a minimum number of training units in the Interim Air Force. Due to this directive, RAAF Station East Sale evolved as a separate Air Force unit on 1 January 1946 under the command of Southern Area Headquarters. The station's former lodger unit, No 1 OTU, ceased to function on the same day and the new unit would command Air Armament School, Air Navigation School and the Crew Conversion Unit (Heavy Bomber). During January, preparations began for the arrival of the Unit's advance parties in the following month; barracks accommodation had been set aside and buildings designated for training purposes were being inspected by the Commanding Officers of the incoming units.

Amazing Statistics

Since starting operations at RAAF Nhill on 8 December 1941, No. 1 OTU had earned a distinguished reputation for training operational aircrew under the Empire Air Training Scheme. The unit relocated to RAAF Station West Sale on 14 February 1942 and moved to RAAF Bairnsdale on 17 June until settling at East Sale. In four years the unit inhabited a quartet of RAAF stations while the instructors trained 3,158 aircrew for active service throughout Australia and the Pacific. Anti-submarine sorties and convoy protection missions had been a part of the unit's flying program which had accumulated a total of 137,962 flying hours by 31 December 1945.

A Vultee Vengeance at East Sale in 1945. Vengeances remained on strength in the RAAF until 1951. (courtesy Reg Adkins)

By February 1946, the mass of discharges dropped the station's strength to only 269 and during the ensuing weeks, the Commanding Officer, Group Captain Deryck Kingwell, the area legal officer and specialist unit officers lectured Air Force discipline, customs of the service, and Air Force law to the remaining personnel to smarten their general standard of discipline and behaviour.

The people of East Sale were delighted with WGCDR Alex Abicair who secured a 100-passenger road train for the station in October. The road train transported station personnel on leave to Melbourne every pay Thursday and returned to East Sale on Sunday evening. The vehicle presented them with an opportunity to escape military life for a weekend and experience the glamour of a cosmopolitan Australian city. The 'train' was a converted semi-trailer towing a white prime-mover fitted with 60 upholstered cinema chairs and a ride cost 10 shillings for a journey to Melbourne which took about 4 hours. The introduction of a reliable transport link out of Sale undoubtedly enlivened the conversations during the quiet working weeks. The RAAF further enhanced the airmens' lives in October with an issue of sheets, pillow slips and mattresses. By November, many chairs were empty in the Airmens' Mess as only 114 airmen worked at East Sale. Overall, a mere 177 personnel manned the station out of a total 'ideal' establishment of 292.

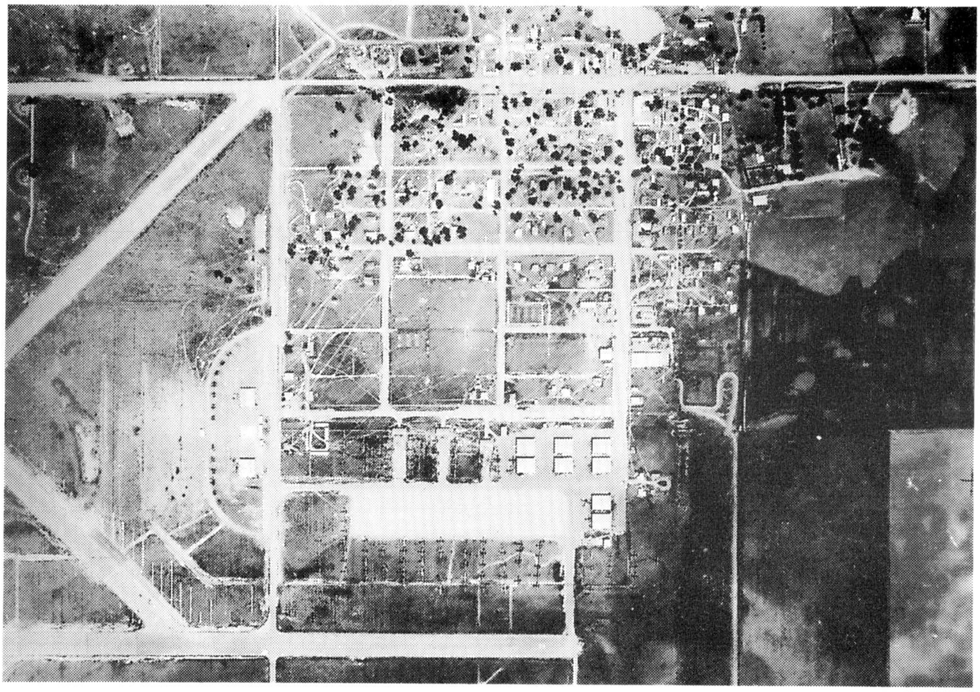

RAAF Station East Sale in 1946. 'Lake Ryland' is at the top of the photograph.

On 22 October 1947, East Sale became embroiled in a local controversy when a Marine Section boat accidentally sunk the tourist vessel Maranui at the entrance of McMillans Strait near Paynesville. The collision gouged a hole along the side of the Maranui which sank to the roof in about 14 feet of water in a few minutes. Three passengers on board the Maranui were injured in the collision and had died when carried out of the sunken boat. Twenty other passengers and two crew members were hauled aboard the RAAF launch and taken to Paynesville for medical attention.

New Construction

After almost 18 months under construction, the 09-270 runway was officially opened on 17 March 1950 when the Country Roads Board Engineer in charge, Mr Dollimore, went for a flight. At a cost of about £120,000, the runway measured 7,000 yards x 150 yards and its construction employed more than 70 Country Roads Board tradesmen. Most of the gravel for the runway came from the Avon River and Stephenson's pit near Dutson. The Country Roads Board development continued at East Sale with the construction of an 8,000-yard runway which employed more than 100 men and about 40 gravel trucks. This extra runway would cater for future intensive operations by bomber, general reconnaissance and jet aircraft. More than 300,000 yards of gravel would fill the runway but this project incurred similar delays to the other runway as the wet weather during 1951 extended its construction by a few years.

On 31 May 1951, East Sale maintained its close association with the district when it participated in the celebrations to herald the proclamation of Sale as a City. The station paraded a guard of honour for the Lieutenant Governor of Victoria, Sir Edmund Herring, who officiated at the ceremony and a Wirraway from Central Flying School dropped leaflets over Sale to publicise the proclamation. Flight Lieutenant Thomas Nichols, the station's armament officer, coordinated a fireworks display on the shores of Lake Guthridge in the evening to complete the day's activities. The first municipality in Gippsland to be declared a 'City', Sale's name honours that of the late General Robert Henry Sale who died at Moodkee in 1845 a hero of the battle of Jalalabad, Afghanistan.

The past two years had ushered in a program of new constructions on the station and with the onset of winter in May 1950, a development scheme focussed on refurbishing living quarters for other ranks. These airmen lived in huts built during the war and each hut housed about 20 personnel. A small cast iron, wood fired stove emitted some heat to the hut's occupants who were issued each with a quilt bearing the RAAF crest, six blankets, sheets and a pillow. As an AC, Bryan Moore shared a hut with his workmates on the station and remembers the evenings warming himself around the stove: 'Cans of spaghetti or baked beans were put on the stove without removing the lids and soon after, the contents of the cans exploded all over the ceiling. We had a friend in catering who was very friendly with the contractor who supplied fruit and vegetables to the station. Once, he managed to obtain a five gallon demijohn of parsnip wine and we had a loud and merry time after drinking a couple of cups of this "bombo juice". The noise became so loud that the duty officer arrived at the doorstep to investigate the commotion. The orderly officer was a Flight Lieutenant pilot, who like other aircrew, considered himself a fine drinker and he disbelieved the "bombo juice" was potent. He downed three cups in succession and then we picked him up off the floor and called the orderly Sergeant to take him away. There were plenty of shocking hangovers the next day and everyone involved was punished with extra duties.'

The interior of an Airmens' Hut. (Courtesy Bryan Moore)

On 30 July 1951, the first intake of National Service Trainees arrived at East Sale to undertake the basic recruit phase of their training. Most trainees spent their time brushing white paint on kerbing and buildings and performing other menial tasks throughout the station. A phrase soon arose that "if you didn't move quickly enough, a National Service Trainee would come along and paint you." During their posting to East Sale, many trainees joined the unit's Australian Rules football team. Their presence unquestionably lifted the on field performance of the RAAF team which defeated the Maffra Rovers in the grand final of the Sale District Football League premiership on 30 August 1952. At least 10 National Service Trainees played in the final and most of those footballers competed in the Victorian Football League, nowadays known as the Australian Football League.

The Orderly sergeant directing the first contingent of National Service Trainees who arrived at Sale on 30 July 1951.

Television personality and motion picture actor Graham Kennedy spent four months at East Sale as a National Service Trainee and recalled his experience in Abby : Portrait of a Common Man: "I hated it all, passionately! I was just getting somewhere with my radio career when the ridiculous interruption occurred. Ridiculous, because the RAAF didn't want us, had nothing for us to do and didn't teach us anything; they didn't even have enough work for their permanent blokes. I chose the Air Force because my father was one of Australia's first air gunners. The Navy would have been a better choice for at least they did something and went places, I just sandpapered silver paint off elderly aircraft. We were told it was Wing Commander Alex Abicair who ordered a silver star to be stencilled on the back of our blue overalls so that if he looked out from his 'tower' he could distinguish between a permanent fellow being idle and one of us being idle - the latter, of course, being okay."

WRAAFs Arrive

Members of the Women's Royal Australian Air Force (WRAAF) joined the National Service Trainees at East Sale when they detrained in Sale on 15 August 1951. During the Second World War women enlisted in the Women's Auxiliary Australian Air Force (WAAAF) and in 1950 the word 'auxiliary' was deleted from the title. King George VI sanctioned the inclusion of 'Royal' in the title so it became the WRAAF and women were now eligible to work in a diverse range of trades throughout the Air Force. The WRAAFs arrival in Sale coincided with the district's highest ever snowfall and the newcomers read about the happenings in the Gippsland Times: "Swirling snowflakes carpeted lawns and gardens, piled up against fences and covered cars. During the height of the storm it was an unusual sight to see pedestrians with their overcoats liberally besprinkled with snow. Men erecting a fence on a district property said the cold was so intense that they had to keep plunging their hands into a tin of hot water to get the numbness from them to handle the posts."

Abysmal weather continued to frustrate the station and the flying program was 'scrubbed' for 10 consecutive days in August. Heavy rains fell over Gippsland for about a week and by 23 August there was extensive flooding across part of the station's perimeter and throughout the surrounding district. However, the postponement of flying in August did not stop the station from accumulating a record total of 13,000 flying hours for 1951 which was the greatest amount of flying East Sale had undertaken since the Second World War. The Commanding Officer, Wing Commander William Townsend, said the figure for 1951 'was by far the greatest total of any station in the RAAF and not an inconsiderable proportion of the total flying hours for the year of the entire Air Force.'

Nearly six months had passed since the first group of WRAAFs arrived at East Sale and on 6 February 1952, the Sale branch of the Air Force Women's Association held a barn dance in the station snack bar to celebrate the first anniversary of the formation of the WRAAF. Barn dancing was fashionable in Sale and Wing Commander Townsend partnered the president and secretary of the association's local branch and mingled among the gathering of airwomen at East Sale. That day became an historic occasion for the Commonwealth when news services announced that His Majesty King George VI had died. The next morning Wing Commander Townsend announced the death of the King at a specially convened parade and among the ranks, officers and warrant officers wore armbands in recognition of his death. The following day, Wing Commander Townsend ordered a parade at midday to announce the accession to the throne of Queen Elizabeth II. On 15 February, the station's personnel paraded again to honour the late King George VI and there was two minutes silence for the assembly which included officers and non-commissioned officers decorated by the late King for gallantry during the Second World War.

Another era ended in April with an announcement that the word 'station' would no longer apply to the Unit after the month. From 1 May onwards, Headquarters RAAF East Sale would exist as a separate Air Force unit under the command of the Air Officer Commanding Southern Area. The unit's Headquarters would exercise direct authority over the Air Armament School, School of Air Navigation, Maintenance Squadron, Base Squadron and the School of Photography. This name change initially gave Wing Commander Townsend a slight administrative quandary: 'The passing of the term "station" has presented somewhat of a problem particularly when one desires to refer to the RAAF establishment at East Sale as a whole. It is difficult to find a word that could replace or surpass the term station when referring to a conglomeration of RAAF functions on the one site. In actual practice it has been found impossible to replace or substitute a suitable term for "station" and the only manner in which one can describe the establishment as a whole is merely to quote the place name, "East Sale".'

The RAAF Women's Association hosted the annual Battle of Britain Ball in the gymnasium which held about 300 guests.

Royal Visitors

A Queen's Colour parade began a series of royal engagements for East Sale between 1953-54. On 1 June 1953, the Commanding Officer of Central Flying School, Wing Commander Charles Read, attended a civic function in the City of Sale Council Chambers to fete the Coronation of Queen Elizabeth II. Coronation revelry resumed in Sale the following morning and amid the festivities, Flight Lieutenant Noel Butler led a flight of four Mustangs over Sale and onto other Gippsland towns between Yarram and Bairnsdale in the afternoon. As a tribute, East Sale held a church service and a general parade where the Officer Commanding, Group Captain William Gibson, explained the significance of the Coronation. A chorus of three cheers for the Queen from everyone concluded the parade and the working day at East Sale ended early that afternoon when the Canteen Service issued free beer to one and all.

Nine months after the coronation celebrations, East Sale personnel had a rare chance to personally see Queen Elizabeth II. On 3 March 1954, the Queen and Duke of Edinburgh deplaned at East Sale and began their tour of Gippsland. Upon the aeroplane's arrival, the Royal Standard broke out at the masthead and the Deputy Premier of Victoria, Mr Galvin, introduced the Queen and Duke to Group Captain Gibson. A procession of Gippsland dignitaries met the Queen before a fleet of 17 black cars from the Royal Car Company drove the Royal Party along the East Sale streets lined by almost 600 service personnel. The motorcade then passed through the base gates onto Aerodrome Road and into Sale. A crowd of about 45,000 people awaited the Royals in Sale and a RAAF contingent of 214 personnel lined the streets to assist a squad of more than 140 civil police. The Gippsland Times described the event as it unfolded: "This was one of the most impressive sights of the tour. With the lovely old Sale Primary School as the background, the street's trees and the continuous rows of children reaching far back into the grounds, there was an indescribable scene. The Queen, a lovely picture, stepped from the car to be greeted by John Cain, the Victorian Premier, and the cheers rang out afresh. Again as the Queen and His Royal Highness mounted the dais, the cheering reached a deafening pitch. Then the band struck up the national anthem. As the Royal Standard broke from the masthead the children and adults along the route took up the anthem to create one of the greatest choirs ever." At the conclusion of the presentations, the Royal Party boarded the Royal train and farewelled Sale to continue their tour of Gippsland.

Queen Elizabeth II and the Duke of Edinburgh departing Sale on board the Royal train.

In 1959, the Victorian Air Training Corps reformed No. 9 Flight at the Sale Technical School after the flight had earlier lodged at Ballarat and Malvern. East Sale had held discussions on the flight's establishment in Sale back in March 1951 and the corps renamed it the No. 9 Sale Town Flight in 1962. By then the flight had a strength of 50 cadets and almost 30 years later, the Mayor of Sale granted the flight, Freedom of Entry to the City of Sale.

The Yankees are Coming!

On 5 May 1959, Gippsland's skies greeted the four fastest aircraft ever to fly in Australia. The 1,200 miles an hour McDonnell RF-101 Voodoo jet aircraft belonged to a United States Air Force contingent of 50 personnel who arrived at the base to participate in Operation Handclasp. The Voodoos had set a world speed record in 1957 and they were the first supersonic aircraft to land in Victoria. A C-130 Hercules, Douglas C-124 Globemaster, SC-54 Skymaster and two Douglas RB-66 Destroyers escorted the detachment of radio servicemen and equipment brought to Australia on a mission described as a goodwill visit. The Destroyers nicknamed, 'Screaming Meanies', created quite a stir upon their arrival. A Gippsland Times reporter described the sound of the aeroplane's engines as a "powerful very high pitched whine" and a RAAF officer reported that "dogs more than a mile away had started to howl as the Destroyers taxied in."

The McDonnell RF101 aircraft on display at the East Sale open day.

This visit became the first of several United States Air Force detachments to East Sale over a 6-year period which would generate great interest among the Australian public. The first task force created favourable impressions among Gippslanders when a base open day attracted 5,000 visitors to view the American aircraft. The United States Air Force would return to East Sale in the next decade but Gippsland people had another opportunity in September to see foreign military jet aircraft. On 21 September, three Royal Air Force Valiant medium bombers and one Comet transport aircraft stopped overnight at East Sale enroute to Edinburgh. The Air Officer Commanding No. 3 Group RAF Bomber Command, Air Vice Marshal M. Dwyer, piloted one of the Valiant bombers to the base where he renewed a wartime friendship with his former navigator Mr D. Shanks. One of the visiting Valiants had recently dropped an atomic bomb during the British tests at Christmas Island and the aircraft's white paint scheme provided an anti-flash protection for this exercise. The Valiants flew to Australia as a public relations tour and had taken part in the Queensland Centenary commemorations.

Sale staged its own celebrations on 22 November when the Council accorded the base, Freedom of Entry to the City of Sale. Mayor Stephenson addressed the parade of 300 RAAF personnel and crowd of several thousand people at the Sale Cricket Ground: 'This is the first time in Australia that the Freedom of a City has been granted to a RAAF unit. My council and the citizens of Sale have a very high regard for the RAAF. We wanted to cement that feeling with the highest honour that we can bestow.' The Mayor then inspected the squadron with the Chief of the Air Staff, Air Vice Marshal Allan Walters and the Officer Commanding East Sale, Group Captain Edwin Fyfe, before a flypast of four Vampires in formation flew over the ground. Superintendent Briggs as City Marshal confronted the RAAF flights as they marched to the intersection of Cunninghame and York streets. Upon acknowledging their right of passage, he allowed the party to march to the Sale Memorial Hall where Group Captain Fyfe dismissed the parade.

The squadron marching along Cunninghame Street in Sale.

As Australia forged friendlier ties with the United States government, developments at East Sale began to reflect this close association. Early in October, a United States Air Force task force of 180 personnel from the 4080th Strategic Reconnaissance Wing of Laughlin Air Force Base at Del Rio, Texas arrived for Operation Crowflight VI. The task force were at East Sale to gather air samples south of Australia for an international study into the rate of fall-out in the upper atmosphere. The United States Defence Department and combined agencies were tabulating this data and the findings would assist a scientific committee of the United Nations.

Major Joe King flew the first air sampling flight on 27 October and piloted the Lockheed WU-2A to within 100 miles of the South Pole. The U2's were fitted with a scoop near the engine intake which contained a particle sampling inlet and a dust screen collected microscopic dust particles. The Task Force Commander, Lieutenant Colonel Kenneth McCaslin, said in RAAF News that 'the U2 was a very delicate aircraft but had no vices. There was always the possibility of flameout because of the altitude we fly at, but that was not a big problem.' The U2's also carried reconnaissance pods and United States Air Force technicians off-loaded the film shortly after the aircraft landed on the tarmac.

The Lockheed WU-2A and Martin RB-57 on public display during Operation Crowflight VI.

A Martin RB-57 aircraft followed the U2's for part of the 7-8 hour sorties. The RB-57 was an American version of the Canberra bomber and belonged to the 4950th Test Group, Kirtland Air Force Base at Albuquerque in New Mexico. Four of these aircraft were deployed at East Sale with two Douglas SC-54 Skymasters. The 76th Air Rescue Squadron of the United States Air Force operated the Skymasters known as the 'Saint Bernards of the Sky'. These aircraft carried on board about two tons of rescue equipment and the Squadron established an airborne radio station and provided a rescue service in support of the operation at East Sale.

The Military Air Transport Squadron of the United States Air Force hauled most of the equipment and personnel across the Pacific Ocean to East Sale in Douglas C-124 Globemasters. A Globemaster then flew a continuous shuttle service between Australia and the United States during the operation.

To coincide with Operation Crowflight VI, East Sale opened to the public on 9 November and almost 5,700 people witnessed a flying and static display featuring aircraft of the United States Air Force. This was the first occasion that a U2 aircraft went on public display and the task force members demonstrated aspects of their work to the crowds. Lieutenant Colonel McCaslin flew the last U2 mission for Operation Crowflight VI on 16 November and the following day a Douglas Globemaster arrived at East Sale to airlift the American personnel back to the United States. Prior to the American's departure, Colonel McCaslin presented the base with a flag of the State of Texas and a Senate of Texas Resolution as a gesture of their friendship.

The apparent success of Operation Crowflight ensured the Americans would return next year but an earlier attempt to undertake this mission had been cancelled. On 7 October 1955, a United States Air Force Strategic Air Command Survey Team visited East Sale in a Globemaster to examine the base and determine if they could mount photographic reconnaissance missions to the Antarctic. The survey team recommended East Sale to the United States Air Force high command but the United States State Department rejected the proposal. The 4080th Strategic Reconnaissance Wing returned to East Sale for Operation Crowflight VII in May 1961. Three U2 aircraft landed on 28 April after a four hour flight from Fiji and its unusual landing technique continued to fascinate onlookers. Major King explained the aircraft's landings to RAAF News during the previous Crowflight Operation: 'I drive a car at speed (near the runway) and talk down the aircraft via radio. In the landing attitude the pilot's vision is restricted and the aircraft is so delicately constructed that it could damage if it dropped in from above two feet. Near the end of the landing run, the U2 which is fitted with tandem wheels located under the centre of the fuselage, drops a wingtip on to the runway. The tips have a vertical projection of extremely hard metal.'

The first U2 flight was on 2 May, one year and a day since a United States Air Force pilot, Gary Powers, ejected out of his U2 after a flameout near Sverdlovsk in the USSR. The significance of that incident and other controversies involving U2 operations in Argentina and Japan elicited a protest by the Victorian Peace Council. The council's Committee Against Foreign Military Bases distributed leaflets questioning the actual intentions of the task force and they rallied a cavalcade of 100 vehicles from Melbourne to demonstrate outside the base. The demonstration had little support from Gippslanders and the protest lost some impetus.

LTCOL 'Buzz' Curry

Operation Crowflight VII continued without further incident and the last air sampling flight finished on 25 May. That evening, a dining-in farewelled the American contingent and the Task Force Commander, Lieutenant Colonel Hayden 'Buzz' Curry, presented the Australians with the Squadron's mascot; a Walt Disney effigy of a crow. Colonel Curry said it was quite a distinction to receive the crow as a base in Fiji was the only other establishment so honoured. The Officers' Mess reciprocated the gesture by giving the task force two kangaroos for homes in the United States. The Americans conducted another U2 operation in May 1963 and then their involvement with East Sale lapsed until February 1965. In that month, 43 personnel and two B-57F aircraft of the No. 58 Weather Reconnaissance Squadron arrived at East Sale to undertake weather research for 90 days.

Ironically, the arrival of the American weather squadron coincided with extreme heatwave weather conditions in Gippsland that led to the worst fires in the area since the devastating grass fires of February 1944. On 3 March 1965, the Sale branch of the Country Fire Authority asked East Sale to provide a ground party and assist the fire fighting operations. Headquarters alerted the base emergency force and a party of 70 personnel and equipment went to Valencia Creek to fight bush fires alight in that area. Meanwhile, a Dakota took off at East Sale for an aerial reconnaissance of the region and these flights extended to two sorties per day during the emergency. To assist the police and Country Fire Authority, a Dakota flew over Gippsland and its crew attempted to spot fires but the prevailing smoke made it impossible to see the ground and the aeroplane diverted to RAAF Laverton as the East Sale airfield had closed due to low visibility caused by the smoke. The first gang of fire fighters from East Sale fought the fires for about two days before a fresh reserve party relieved them. The base operations room, duty crew, firemen and motor transport drivers provided 24-hour assistance until the crisis ended on 12 March when light rain reduced the fires intensity. A team of up to 80 RAAF firefighters were in the field during the emergency and their efforts probably stopped the fire engulfing the small town of Newry. A Gippsland Times reporter entered the fire sweeping across Glenmaggie on 3 March and watched the 'flames 50 to 100 feet high' moving 8 to 10 miles an hour. Stock crashed through fences to safety. Police and firemen fought against choking smoke and on hill tops, you had to lean against the wind so as not to be blown down,' he said. Bushfires also started at Briagolong, Stockdale and Fernbank and farmers reported stock losses amounting to several hundred sheep and cattle as the fire spread over a million acres of Gippsland countryside, destroying almost 40 houses along its path.

On 9 April 1970, East Sale resumed its close association with Queen Elizabeth II. A Royal Australian Navy Iroquois helicopter of No. 723 Squadron flew from East Sale to deliver the Royal Mail to HMAS Britannia, 30 miles off the Gippsland coast. Exchange officer Lieutenant Gerald Hardy of the Royal Navy captained the helicopter which winched the mail down to the Britannia and to HMAS Stuart, an Australian destroyer escorting the Royal Yacht. The Crown Government Postmaster had previously arranged the delivery of the Royal Mail to Sydney via the United States. The Base also provided a search and rescue commitment for the Britannia during its voyage from Sydney to Melbourne.

The RAAF continued its aid to the Gippsland community on 4-5 November 1977 when East Sale deployed men and equipment to Morwell to fight a fire in the Morwell open-cut brown coal mine. The Morwell police made an urgent request for help and within an hour 80 men and three fire tenders were heading to the mine operated by the State Electricity Commission. The party split into two flights of 36 men who began extinguishing fires in the mine on level five and on the ground floor. RAAF News described the scene: 'The fire at one stage was burning on a 1 kilometre front with smoke billowing as high as 350 feet. It appeared like a giant radiator as it burnt down the full 150-metre face of the open-cut mine.' Squadron Leader Larry Crowley led the Special Emergency Services and wrote a report about the fire: 'At level five the problems were worst because of intense heat, smoke and cinders. The vehicles were moved off in a hurry because small fires were breaking out everywhere, including under the vehicles. These two flights then attacked the small fires with their limited equipment and maintained a clearway for vehicles moving through. At level eight, the other two flights were dispersed along the ridges to the north and east; again because of equipment limitations, the fighting effort was only a containing exercise. Once again, the vehicles were returned because of the fire threat. As more hoses became available, the men were able to make a worthwhile contribution, attacking the ridges above and below them. After some disappointing wind changes, where hours of work were lost in seconds, the wind abated and it rained. By this stage, most of the men were thoroughly wet and very cold.' The RAAF flights left the coal face after midnight and slept in guest houses in Yallourn near the mine. They continued combating the fire shortly after sunrise and continued till midday when a reserve flight of airmen from East Sale replaced them. Late that afternoon Squadron Leader Crowley withdrew the men except the professional East Sale firecrews who stayed until the following morning.

For more than 30 years, East Sale had helped Gippslanders overcome natural disasters and the base still resembled the RAAF station which had accommodated No. 1 OTU during the Second World War. Many buildings built then were still in existence at the base in 1978 but a development scheme for East Sale would begin a modernisation program that would extend into the 1990's. In 1978, the RAAF coordinated the construction of: a $502,828 precision measuring equipment laboratory; a new $345,000 barracks service complex; renovations to the airmens' kitchen costing $320,000; a $140,000 Macchi flight line building and new annexes and walkways between aircraft hangars costing $142,000. The RAAF had also negotiated a $656,259 contract with A.V. Jennings for the erection of 21 new houses to replace the old dwellings on the base.

As East Sale entered the 1980's, the base assisted the trial of new fire fighting equipment. In January 1982, East Sale was a mixing and loading base for the Modular Airborne Fire Fighting System in Victoria. A Hercules from RAAF Richmond airlifted the equipment and conducted trials over bushland in East Gippsland. The previous year, the American Forests Services had trained RAAF personnel, forestry rangers and a CSIRO technician to use the system and two American specialists supervised the tests. The system contained 13,600 litres of Amgard, but water replaced the mixture for the experiments. The Minister for Primary Industries and Member for Gippsland, Peter Nixon, said the system 'would enable the Victorian government to utilise a technologically advanced bombing system in the fight against the bushfire menace in Victoria.' The trials became the forerunner for further bush fire tests across East Gippsland in 1984. Code named 'Project Aquarius', the CSIRO and East Sale scheduled the trial in early January till late March. The project evaluated the technical and cost effectiveness of large air tankers for the control of fires in eucalypt forests throughout Australia in comparison with conventional fire preventative measures. The trials engaged a Douglas DC-6B on lease from Conair Aviation of British Colombia and it airlifted on separate occasions about 11,000 litres of water, thickened water and retardant. The aeroplane dropped these mixtures over 16 intentionally lit, 100-hectare blocks in the Nowa-Nowa forest. Each block contained fires of differing intensities and the Chisholm Institute of Technology gathered the information from these tests and applied the data to a mathematical model.

On 17 November, RAAF East Sale contributed to Victoria's 150th anniversary celebrations with a march through Sale to exercise its right of entry. The 4/19th Prince of Wales Light Horse Regiment followed the four East Sale squadrons along the city streets and No. 1 Flight of the Escort Squadron assembled in Raglan Street for a dedication ceremony. The Officer Commanding, Group Captain Allan Pappin, presented a mounted HS748 propeller as a RAAF memorial, to the Mayor of Sale, Councillor Geoff Rossetti. This was Sale's first event to celebrate Victoria's historic anniversary.

The year of 1988 heralded the Australian Bicentennial and East Sale marked this notable occasion in February with an official opening ceremony for the Avenue of Honour and a base open day. The Avenue of Honour scheme started in May 1987 and employed 16 Sale people with assistance from East Sale, Sale City Council, Avon Shire, Sale Rotary Club, Stratford and Sale Lions Clubs, Sale's Commonwealth Employment Service and Stratford and Sale Returned Services League Clubs. These bodies created the avenue to recognise people from the district who served Australia in war between 1899-1972. Memorial plaques were placed at the base of each of the 500 trees planted along both sides of Aerodrome Road and each plaque commemorated three persons who served. The Sale community financed the project and Flight Lieutenant Dave Stockdale and Flight Sergeant Cam Wilson of Facilities Flight provided specialist advice. The base officially opened the Avenue of Honour on 18 February and since then East Gippsland's No. 9 Flight Victorian Air Training Corps has helped to maintain this picturesque tree lined path leading to East Sale.

This event preceded East Sale's family day on 20 February and the public open day the following morning. About 32,000 people attended the open day to walk around RAAF aircraft static displays, see fire fighting and police dog handling demonstrations while listening to the RAAF Central Band. People gazed skywards to see the spectacular Roulette High Show Display and solo displays by other contemporary RAAF aircraft, historic RAAF Museum aeroplanes and Navy aircraft.

On 21 October in the Bicentennial year, the Minister for Defence, Kim Beazley, officially opened the East Sale development works scheme. Divided into Phase A and B, the immense construction program started at the base in May 1986 and cost about $25 million. Phase A involved the erection of junior and senior officer accommodation, the School of Air Traffic Control, Central Flying School, Maintenance Squadron and Field Training Flight, Base Security Control Centre and the School of Photography. The second phase started on 26 November 1987 and embraced the construction of the other ranks accommodation, Fire Station, School of Air Navigation, Communications and Electronic Data Processing Centre, Security Guard/Police Dog Complex and landscaping across the base perimeter.

The development financially benefited and encouraged the local district. The Administrative Staff Officer, Wing Commander Rick Phillips, said in the Gippsland Times that 'the construction of the new facilities would give the local community confidence that the RAAF would remain as a significant employer and source of income for many years to come.'

This works project immeasurably enhanced the beauty of East Sale and helped it to win the 1988 Military Bases Section of the Keep Australia Beautiful, Tidy Towns Competition. Tidy Towns judge John Ernst inspected East Sale in September and said he admired the community activities undertaken by the base, its 'Ali Can' Cans recycling program and the attractive landscaping around the new buildings which aesthetically complemented the surrounding environment. The Premier of Victoria, John Cain, presented the Tidy Towns award to East Sale on 2 December 1989 before 700 guests at a ceremony on the base. The guests included numerous town and shire representatives, senior military officers of the Australian Defence Force and the Air Officer Commanding Support Command, Air Vice Marshal Alan Reed, who presented the Defence Force Award. The RAAF Central Band played a musical accompaniment and afterwards the guests watched a Roulette display. In 1992, the Base again won the coveted award.

East Sale's development program extended into the 1990s and a RAAF defence housing and redevelopment project further improved the lives of people posted to the base. In early 1990, the Australian Federal Government allocated $750 million for the removal of substandard defence housing across Australia and Sale would get about $1.5 million. The Defence Housing Authority had purchased about 15 sites in Sale designated for new married quarters and the authority would increase its ownership of local homes from 326 to almost 430. To achieve this aim, the authority intended to return 56 houses to the Victorian Government and sell other inferior housing and then buy and erect residences of an appropriate standard. Sale's building contractors would also benefit from the project as the authority intended to subcontract maintenance work on the houses. East Sale DHA manager Mr John Seymour now proudly boasts the highest customer satisfaction in mainland Australia.

In April 1990, East Sale registered its fifth decade of aid to Gippsland authorities when the base emergency force provided flood relief assistance. On 20-21 April, the meteorological section tallied record levels of rainfall in the Thomson, Macalister, Avon and Mitchell rivers which caused flood waters to rise in the upper tributaries of the Avon river at a rate of one metre an hour. On 21 April, the State Emergency Service and the Victorian police asked East Sale to assist their flood relief operations. In response, the base emergency force assembled 30 personnel who manned road blocks and evacuated people out of flood bound houses for two days. After the flood waters eased, East Sale donated bedding and cleaning supplies to the Community Services of Victoria and HS748 and CT-4 patrols evaluated the extent of flooding across the region for the police and the State Emergency Service.

On 23 September 1992, the Managing Director of Hawker de Havilland, Bruce Hattersley, presented the Hawker Siddeley Trophy for 1991 to the Officer Commanding East Sale, Group Captain Peter McDermott at a specially convened parade of base personnel. The Chief of the Air Staff, Air Marshal Ray Funnell, awarded the trophy to East Sale for the quality of training and efficient administration prevalent on the base and for the fine dress and bearing of its personnel. The Hawker Siddeley Trophy represents an annual competition between all the RAAF bases and East Sale had previously won in 1972 and 1985.

CHAPTER 7

MAINTENANCE SQUADRON EAST SALE

by Flight Lieutenant Peter Hayward

On 1 May 1952, Maintenance Squadron East Sale (MNTESL) was formed from the technical component of RAAF Station East Sale, with Wing Commander A.J. Abicair, MBE (now deceased) as its first Commanding Officer. At the time of its formation MNTESL was responsible for the maintenance of 10 different aircraft types, these being the Lincoln, Dakota, Mustang, Wirraway, Vampire, Beaufighter, Tiger Moth, Mosquito, Chipmunk and Auster.

Snakes, Rats and Royal Visits

During its early period of operations, the Unit faced many difficulties; initially the unit strength of six officers, 251 airmen and two WRAAFs was well below the establishment of ten officers, 313 airmen and five WRAAFs. Combine this with the operation of the ten aircraft types and, at times, severe weather conditions, resources were severely strained as these two entries indicate:

> '11 July 1952. Bad weather conditions for Maintenance Squadron; 16 days of rain, seven days when frost occurred, made the maintenance of Unit aircraft, the majority of which were parked in the open, an unpleasant and difficult task.'

> 'Late 1952. With the advent of warmer weather, snakes are appearing on the aerodrome, but to date have given no cause for concern. Rats which prove troublesome by eating electrical wiring in aircraft have been exterminated by the laying of poisonous baits.'

But even in the first year or so of independent operation, time was found to demonstrate loyalty to the monarchy, contribute to the development of the Base and to forge strong links with the community. This was evident during the commemorative church service and parade held to mark the coronation of Queen Elizabeth II in June 1953.

1954 saw the introduction of a new internal organisation considered to best meet the demands placed upon the Squadron. The unit was divided into sections of heavy aircraft, light aircraft, armament and workshops, ancillary section, tarmac, records and Ground Equipment Support Maintenance (GESM). Notwithstanding continuing poor working and weather conditions, 1954 saw a post war record flying effort achieved with 11,916 hours being flown on all aircraft types. This was followed the next year by a unit record monthly maintenance effort of 2,240 engine flying hours in conjunction with 990 airframe flying hours.

1954 also marked the establishment of the Jet Training Flight:

> '57 personnel of various technical musterings arrived on attachment and comprised the maintenance section of the Jet Training Flight. This experimental section officially began functioning on the first of the month although flying did not commence until the 10th of March. The aim of Jet Training Flight is the conversion of pilots direct from basic flying training onto Vampire jets.' Also that year, newly crowned Queen Elizabeth II and the Duke of Edinburgh were able to fit Sale into their Royal Tour, landing at the Base and travelling into Sale.'
>
> 'Her Majesty Queen Elizabeth II and the Duke of Edinburgh accompanied by officials of the royal tour landed by aircraft at RAAF East Sale at 1135 hours. After brief presentations on the tarmac she drove by car to Sale. The route of the royal progress on the formation was lined by all ranks including all the airmen, NCOs and officers of this squadron. The turn out of this Squadron's personnel was extremely good and was favourably commented upon by the Officer Commanding.'

1955 saw the move of Tiger Moth flying from East Sale to its satellite base at West Sale. Although the aircraft only flew every second day to allow for aircraft repairs, all the 'petrol, oil, spares and handling equipment' had to be transported to the West Sale aerodrome from East Sale. An additional WOFF and three airmen were also diverted to West Sale on these flying days. September of the same year saw the introduction of Winjeel aircraft into service at East Sale.

Late 1958 was noted for the sad end to what was thought to be a prank. In the early hours of a November morning a LAC engine mechanic from MNTESL believed in his abilities to fly a Dakota aircraft. The aircraft, A65 89, crashed on take off and was destroyed by fire with the would be pilot being instantly killed.

Engine Fitter at work on a Lincoln

New Decade, New Types

The passage of the 1950s saw a phasing out of service at East Sale of the Lincoln, Mustang, Wirraway, Beaufighter, Tiger Moth, Chipmunk and Auster aircraft. During 1967 and 1968 the aging Vampire was replaced by the Macchi, which is still in service today although no longer operated on the Base. Soon after their introduction, the first loss was recorded on the 7th of March 1968 when aircraft A7-003 was destroyed in an aircraft accident. Also that year the first HS748 Navigation Trainer, A10-601, was received into the RAAF at East Sale as work commenced on two hangars with the anticipation they 'be completed before the first HS748 is due for "D" servicing'. Aircraft A10-601 and the remaining aircraft were received over the next 12 months, aircraft numbers A10-602 to 608, are still in service at East Sale today.

A Take-over Bid?

Whilst Maintenance Squadron had, since its inception, serviced ground support equipment such as tow motors and power carts, it was not until 1969 that an authority was received to transfer mechanical transport maintenance (cars, trucks, fire tenders etc) from Base Squadron. While this responsibility has remained to this day, it will be the subject of the Tier 2 of the Commercial Support Program currently being evaluated at East Sale.

April 1971 marked the reorganisation of the Squadron into a series of Flights and Sections very similar to today's structure. On the direction of the Department of Air, a new organisation of Maintenance Squadron was implemented. The Squadron was divided into flights and sections as follows: Maintenance Support Flight composed of MCS and Equipment; Field Training Section composed of Training Control Office, Technical Training School and Trade Test Office; Ground Equipment Maintenance Flight composed of Ground Support Equipment Maintenance, MT Maintenance and General Engineering; Aircraft Maintenance Flight composed of A Flight Line (Macchi, Dakota, Winjeel), Macchi Maintenance, Dakota Maintenance and Winjeel Maintenance Sections, B Flight Line (HS748) and HS 748 Maintenance; Aircraft Equipment Maintenance Flight composed of Safety Equipment, Radio, Electrical, Instrument and Armament Sections; and Administration Flight composed of the Orderly Room and the Disciplinary Office.

1972 marked the 'crisis point' for the old Safety Equipment building. Reports show that four years previously the wartime constructed temporary building had to be winched back to the vertical position, propped and stayed, after it began to fall over during a time of high wind. By 1972 nothing had been done to replace the building and in fact the planned replacement had yet again been dropped from the New Works Program. This prompted a visit from the Works Architect who was scathing in his report:

> 'It is my considered opinion that this Safety Equipment Building is possibly the worst example of sub-standard and unsafe accommodation at present in use throughout the RAAF. It is also my opinion that until such time as it is replaced, an instruction should be forwarded to the Base to state clearly that the building is not to be occupied during periods when the wind strength exceeds 40 knots.'

The building was finally replaced in May 1975 following considerable controversy over whether the building met the stipulated standards.

Concern for the Future

1973 saw an occurrence which today could be the reason for a feeling of deja vu. Unit history for July 1973 shows uncertainty and unrest becoming apparent over disclosures of possible reductions in the strength of the RAAF. Commanding Officer at the time, Wing Commander B.J. Graf, sought to allay people's fears by suggesting that all was just rumour until the budget was brought down in late August. Even then, he said, 'large scale reductions are not anticipated.' In September however, their worst fears were realised:

> 'Morale of the squadron was adversely affected when the extent of the establishment reductions recently received at the unit became known. Of particular concern was the reduction in the officers and SNCO establishments and with this the supervisory capacity of the squadron. Typical of the reductions was the elimination of five WOFF posts of the eight previously established. In fact, 16 of the 42 posts disestablished were posts of SGT

and above. Fortunately as yet no large scale moves as a result of these changes have been made, but this has not eased the general feeling of uncertainty generated by the establishment redutions. Moreover, an increase in applications for discharge under the new relaxed policy has been evident.'

The same year that MNTESL supported Papua New Guinea's celebration of Independence and the Winjeel aircraft departed from the base for the last time. The Winjeel had served admirably as the primary flying trainer over many years.

In November 1979 Macchi Flight line personnel vacated their World War II flight line building and took up residence in temporary accommodation. This enabled the old flight line building to be moved from the Macchi Hangar area to a temporary position in front of the Dakota Hangar. The temporary accommodation for flight line personnel was a large marquee erected on the tarmac area between the Macchi 'coat hangars'. Unfortunately 40 knot winds caused problems with the flight line tent resulting in 12 concrete aircraft tiedown weights and numerous steel bars to eventually secure the tent.

Temporary Macchi Flight Line hut

When the old flight line hut was eventually lifted and removed from where it had stood for over 30 years some interesting relics that had been left there over the years were revealed. Included among the relics were a 1944 'FLIGHT' magazine, a 1954 'ARGUS' newspaper, assortments of old cigarette and match boxes, pre 1960 beer bottles and empty ammunition shells believed to be from Meteor jets.

Early in 1980 the Dakota aircraft, with many a tear in the eye of those that flew or maintained her, departed the Base for the last time. This feeling of pride was only dented once when in 1975 for reasons unexplained, a propeller departed from the aircraft and with near t ragic consequences, struck the side of the aircraft narrowly missing the person sitting at the Navigator station. A positive outcome of the incident was the practical experience gained by the Aircraft Metal Workers Course that travelled to East Sale to effect repairs to the aircraft as part of their practical phase. Needless to say the repairs were of the highest standard and the students passed.

Ceremony to mark the sad departure of the Dakota aircraft

In 1985 FSGT John Mowbray organised a successful function for 120 past and present DC-3 pilots and technicians to celebrate the 50th anniversary of the first flight of a DC-3. The function was held in the Airmen's Mess with ARDU supplying a DC-3 for a static display.

50th Anniversary Reunion Function
Approximately 120 past and present RAAF DC3 pilots and technicians attended a function at RAAF Base East Sale on 17th December 1985 for the 50th anniversary of the first flight of the DC3 aircraft. A reunion luncheon was held in the Airmen's Club Function Room to commemorate the outstanding service this classic aircraft gave the RAAF. ARDU kindly provided a static display of their DC3 on the day.

Only months later, Hawker De Havilland were awarded the initial maintenance contract on the recently delivered PC9 aircraft. Several serving personnel took this opportunity to join 'Hawkers' by trading in their uniform for civilian attire while still satisfying their desire to work on the PC9 aircraft. At the time, considerable unrest was noted amongst the technical airmen who were now unable to use their skills on some of the 'new technology'. Although Hawker De Havilland no longer hold the contract, most of the ex-serving personnel are still at East Sale working on R4 and R5 servicings of the HS748 fleet.

Records Set

Also in 1986, a milestone was reached. A young pilot and a venerable old bird teamed up to register the 10,000th flying hour for a RAAF HS748. Aircraft number A10-601 was accepted by the RAAF in London on 6 September 1968, and since arriving at East Sale has been professionally maintained by the Squadron. During its 10,000 flying hours the aircraft has flown over most parts of Australia during navigation training exercises. A10-601 has undertaken several other tasks outside its prime role including flying the King of Nepal and being a back-up aircraft for the visit by Prince Charles and Lady Diana in 1985. Four years later, the HS748 Navigation Trainer aircraft fleet maintained by the Squadron recorded a total of 100,000 accident free flying hours - a milestone in anyone's book.

Early in 1988 officers of the Squadron conducted their own bicentennial re-enactment of 'Transportation' when an ex-RAF Engineering Officer Flight Lieutenant 'Mac' Robertson was recruited and posted directly to OIC AMS. To record the re-enactment, the Commanding Officer despatched an appropriate message to Air Force Office postings staff. Mac was presented with his 'initial uniform' and a copy of the message as a memento.

By October 1988 the new Headquarters of the Squadron was completed with guest of honour at the opening ceremony being Air Commodore B.J. Graf, a former CO of the Squadron. Six weeks later the Squadron was again in the limelight when it was presented with the RAAF Maintenance Trophy for the 87/88 year.

Governor General's Banner

The highlight of 1989 was the presentation of the Governor General's Banner to the Squadron. Given in recognition of 25 years outstanding service by a non-operational squadron, the banner was presented with due ceremony by the Governor-General Mr Bill Hayden, AC. In a moving ceremony, the sky blue flag with wattle, fringed and tasselled with blue and gold, was handed over to the then Commanding Officer, Wing Commander G.F. Bates.

Governor General presenting MNTESL with their banner

Civilianisation

January 1 1991 marked the expiry of the civilian contract for PC9 servicing. As a result, 16 of the civilians filled newly created positions at HS748 Section and were given responsibility for R5 servicings. By the end of June, flying hours for the 90/91 financial year totalled 13,764. On the 10th of December that year all ten CT-4 aircraft departed the Base for the last time. Among them was aircraft number A19-048 which had been classified as having Category 4 damage from a crash earlier in the year. Members of the Squadron were able to restore the aircraft to a serviceable state and allow it to rejoin the fleet.

The high side of aircraft operations

Maintenance Ingenuity and Innovation

November 1955 - Maintenance Squadron has always been at the forefront of innovation and ingenuity willing to try anything new. Consequently, recorded in unit history for the time was a visit by Mr Lovie of the Organisation and Methods Section at the Department of Air who had come down to investigate a Maintenance Squadron proposal. Record Section had 'stirred the possum' by placing a serious request for approval to use an adding machine in the workplace. Now that's what you call ambitious use of new technology!

March 1959 - The limelight was again on the Squadron with its 'leading edge' experimental use of luminous orange paint being applied to aircraft to minimise the possibility of collision. Needless to say the orange paint became standard livery for training aircraft.

June 1964 - The initiative of an alert young airframe fitter was responsible for the detection of fatigue failure of tailplane attachment bolts in the Winjeel aircraft. Leading Aircraftsman J.W. Newman was carrying out a "C" servicing when he noticed excessive movement of the tailplane. On checking he found the tension of the nuts to be unusual. Further checking revealed one bolt was sheared completely and the other was cracked in two places. The only other two bolts holding the tailplane at the forward end showed wear and fret marks. As a result Winjeels were grounded and fatigue was later confirmed as the cause of the damage.

July 1965 - A special committee was formed to investigate the increasing incidence of foreign objects left in aircraft. Items found included tools, bolts, washers, screws, money and even flying boots. In direct response to the discovery of tools left in the serviced aircraft, Maintenance Squadron introduced the concept and practice of composite tool boards for servicing bays.

May 1972 - Leading Aircraftsman A. Van Diem, an instrument fitter, observed that external rescue markings "Cut Here In Emergency" on an HS748 were in an extremely dangerous position. A high pressure oxygen pipe located behind the sign would have been severed had the aircraft skin been cut at this point. This would have released oxygen at 400 psi with potentially very serious consequences.

February 1986 - Motor Transport spares had historically been purchased from Melbourne based period contractors. However, the consequential transport cost and delays far outweighed the on-cost of purchasing spares from the Sale area. Headquarters Support Command finally agreed to the Squadron's request for local area Motor Transport (MT) and Ground Support Equipment (GSE) spares purchase. Even after two weeks of operation MT and GSE down time had improved.

Forty Years Young

Today, forty one years after commencing operations, Maintenance Squadron reflects the steady progress that has been made since its formation. Most of the old war-time galvanised iron or fibro-cement clad buildings have been replaced with permanent brick buildings, two large hangars have been constructed for the maintenance of HS748 aircraft, two steel structures on the flight line now provide shade and shelter for PC9 aircraft and hangar facilities exist for the servicing of the PC9.

With a unit strength of 220, the Squadron provides and maintains 16 PC9 aircraft to Central Flying School for instructor pilot training, conversion and refresher flying, and for the Roulettes Aerobatic Team. Ten HS748 aircraft are maintained for use by No. 32 Squadron for Navigator and RAN Observer training and for general transport tasks for the ADF. While the unit fully maintains a large number of avionic items, the more complex communication items are despatched to centralised facilities for repair and overhaul. Responsibility is also held for the servicing of all motor transport vehicles and ground support equipment in use on base. In addition, Squadron personnel maintain flying safety equipment, provide fuel quality control services for the fuel installations and service the fuel tankers in use. The Squadron also has responsibility for the calibration of all test equipment used on the Base.

But perhaps the greatest challenge is still to be faced by the Unit. Identified as part of the Commercial Support Program's Tier 1 activities, the Unit is undergoing considerable restructure to enable it to be competitive against the private tenderers seeking to assume the primary role. The Squadron will continue to progress but perhaps in a vastly different way compared to previous times. Less uniformed people will certainly be one outcome with more of the support tasks being done by civilian employees or by private contractors. Whatever the final outcome, the spirit of camaraderie, friendship and co-operation will continue to be the backbone of the unit.

A rare but welcome sight - all aircraft serviceable and in the air

CHAPTER 8

THE SCHOOL OF PHOTOGRAPHY

by Chief Petty Officer Steve Given

Early Courses

The RAAF's first photographic Unit began as a flight originating from a wartime nucleus of photographic personnel at Point Cook, Victoria in 1941. The flight later transferred to RAAF Canberra as part of a Survey Flight in 1943. Twenty seven courses were conducted for airmen at Canberra up to the end of the war in 1945. The Photographic Flight was reactivated after a post war lapse and transferred to RAAF East Sale in October 1946. Commencing with Number 28 Course, the new School offered tuition to new intakes and later to General Duties Officers courses. A basic course of three months, and a secondary course of nine months were firmly established at this time with the first course being conducted by Flight Lieutenant D.J. Robertson.

Students were instructed in basic photography, chemistry, optics, processing and photographic equipment, and used the S1 (half-plate wooden), S4 (whole-plate metal) and C3 (4x5 Speed Graphic) plate cameras. These courses were comparatively short compared with those of today, although a secondary course was conducted at a later date. From the first course, photographers were posted for duty to No. 87 Squadron (Mosquito Aircraft), Canberra ACT; No. 81 Wing, Japan; and Central Photographic Establishment (CPE) at Laverton, Victoria.

The 1950's - Courses Begin

The School of Photography (PHOTS) was formed from the Photographic Flight, Headquarters East Sale on 2 April 1952 and was to function as a separate Air Unit at RAAF Station East Sale under the direct command of the Commanding Officer RAAF East Sale. Flight Lieutenant R. H. Campbell was posted in to become the first Commanding Officer of the new school, but was posted to RAAF Headquarters on 5 May in the same year. On 2 May, Flight Lieutenant Kip Porteous was posted in to command the school and remained in this position until 19 July 1954. During that time, Basic Airmen's Courses were run concurrently with emphasis being placed more on practical work than theory.

No. 1 GD Officers Photographic Course - 1949

No. 1 Senior NCO Course - 1954

As the commanding officer was the only officer on the Unit, an inordinate amount of time was taken up with administrative matters. There was no officer to take over even in the event of hospitalisation. This situation was experienced by Warrant Officer C.G. Smith, who at one stage had to hold down the jobs of chief instructor, adjutant and warrant officer disciplinary.

During the very early stages of the formation of the School, efforts were made to procure more buildings. Most of the available buildings would have required alterations not commensurate with the return and it was not until August of 1952 that the Unit acquired a building which was deemed as suitable for the School Headquarters. This building was taken over during September and at the same time to the delight of the staff, new equipment arrived at the school. This was also the time that the students nemesis, the lawns and shrubs which required constant attention, were first planted, and the gardens started. This 'extra duty' was to be later recognised during the graduation of No. 41 Basic Photo course where all members were also presented with Diplomas from the RAAF School of Agriculture!

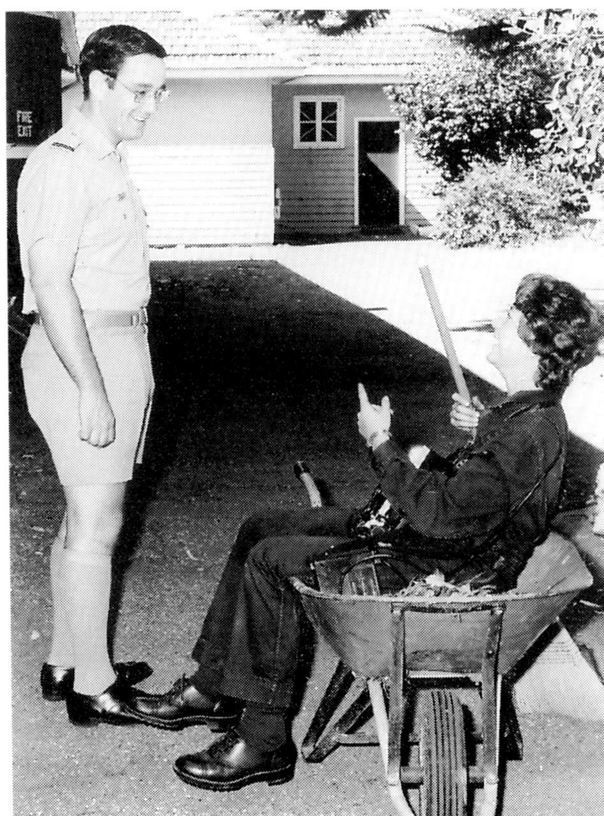

Chief Instructor, Flying Officer 'Wally' Walton and a reluctant student gardener, ACW Tina Snow

Training of No. 9 Navigation Course commenced at the new School in October 1952, with instruction covering 'Hand Held Oblique Photography'. By the end of 1952, the standard of equipment and buildings was reported to be deplorably low and it was difficult for staff and students to achieve good photography while working under conditions, using primitive equipment.

Full scale instruction of photography to Officers, NCOs and Airmen began in 1953 with three courses being held during the year. The courses were reported to have had their share of 'diverse personalities' and rank levels, which ranged from Aircraftsman through to Flight Lieutenant including a Pilot Officer with a DFM.

In 1956 the first Adjutant, Pilot Officer K.J. Wilson, arrived on strength at the School. In December of 1957 the Unit received a commendation from the Air Officer Commanding, Air Vice Marshal I.D. McLachlan, for efforts in producing recognition material for the three Services.

A Place to Call Home

Overdue renovations were carried out on a number of the acquired buildings which were being used as classrooms and darkrooms. Their condition was described as 'a little short of sordid.' These renovations did little to help with the manning and training schedules, which resulted in course overlapping due to suspended practical training. Also acquired during 1953 was another HQ building, and two lecture buildings with one of the latter being equipped as a studio. In the last half of 1954 a sensitometry lab was set up to raise the standards of photo training, and carry out tests on new films entering the Service. Also the first Senior NCO Refresher Course commenced in October of that year.

The Old School which eventually became the Store and Darkrooms

New acquisitions for the year of 1955 were new enlarging equipment, and another building for the use as a chemical and glazing room. The communal laboratory (Comlab) was used for the first time for No. 44 Basic Course. In May, the Australian Paper Mills Trophy was presented to the School and awarded to the dux of the GD Officers course. In June, Warrant Officer C.G. Smith, who had been the Unit Warrant Officer since it's inception, was awarded an MBE in the Queens Birthday Honours List.

The Badge

The official RAAF School of Photography Badge was registered at the College of Arms in March 1955. The Badge depicts a torch, the traditional symbol of learning, on a Sun, indicative of the light that is essential to photography, against a white background. The Motto of the School is 'EX LUCE DISCIMUS' which may be literally translated as 'We Learn from Light.'

An Unofficial Badge - Tongue in Cheek!

The 1960's

Equipment shortages were apparent in mid 1960, when there was only enough stocks held for 12 of the 16 members on course. It was at this time that the first discussions were held for the proposed amalgamation of PHOTS and CPE. An Instruments Workshop was also established in 1960. On 4 October 1961, a meeting was attended at CPE to discuss the proposed plans for extensions to that Unit to accommodate PHOTS.

As the Fifties were replaced by a new decade, so were many of the archaic photographic materials and processes. In February 1962 the School was directed to carry out tests on Kodak Plus-X Aerographic material, a 'new' aerial film being produced in Australia. In the following month, staff and course members were lectured on the introduction of new materials such as colour print materials and equipment. Later in the year, the School of Photography starred in a television program 'This is Gippsland', produced by a team from the local TV channel (GLV10) A further boost to morale occurred when the staff and students took receipt of the first six new Linhof Super Technica cameras.

By mid 1964, outfitting of additional darkrooms in Building 379 at the School brought the number of training darkrooms to seven. Approval for the new Airmen's Course syllabus was also given in April, and No. 53 Airmen's Course was the first to use the new syllabus where practical work was introduced during the first week of training. The Linhof Trophy which the School won twice in three years, was presented to PHOTS to encourage and develop skills of RAAF photographers throughout the world.

Disastrous fires swept through Gippsland in March 1965, and members of the School of Photography, including staff and students, fought fires at the Tambo Crossing and Newry areas.

By the end of this year the last of the equipment for a colour capability at the School was installed. No. 57 Airmen's Course was the first course at the School to receive practical training in colour photography along with Nikon F cameras which were also new additions at this time.

Bob Hawley, prominent School Instructor in a Dak in 1977

In July 1969 as part of the increased emphasis placed upon quality in aerial imagery, the School introduced a Quality Control Course. Also in mid year, aerial photo training shifted from the Dakota aircraft to the HS748 Aircraft and No. 14 Officers Course was the first to use the new aircraft for such training.

At the end of the decade, PHOTS consisted of seven buildings, all of wartime vintage, and all generally substandard for training purposes. The buildings housed nine darkrooms for training, four for staff work, and one for trials, together with studios and workrooms. The staff of the School consisted of 20 officers and enlisted men including the Commanding Officer, Squadron Leader Riggs, and the Chief Instructor, Flight Lieutenant Wilson. The instructional staff comprised a Warrant Officer, a Flight Sergeant and two Sergeants. There was also a staff section of one Sergeant, one Corporal and three LACs, a Trials Section with a Flight Sergeant and a Corporal, and an Administrative and Stores staff each of two people. A camera technician and a general hand comprised the remainder of the staff.

The 1970's

On 18 March 1970, after many years of discussions, plans for a new self contained building to house the School were submitted for approval. Little did the submission originators know that it would be nineteen years before they would be able to 'wet the new baby's head'.

In 1971, possibly as an appeasement at not getting a new School building, the floor boards of most buildings were layered with continuous vinyl to replace the worn out lino. Supplies of materials dwindled to a bare minimum mid year due to Government financial restrictions. During the period, after pushing for materials, the School received an inquiry requesting reasons for demands. In reply, the CO wrote 'No Supplies; No Training; End of Story.' An Instrument Cleaning Machine and Zelox Electronic Camera Tester arrived in August, and was said to 'gladden the heart' of the Instrument Fitter, who previously had to borrow equipment, or use 'primitive trial and error methods'.

During 1972 there were proposed plans to have an integrated school building to house a combined Central Flying School (CFS), School of Air Navigation (SAN), and PHOTS. However, after many discussions, the proposal was dropped as it was found to be not in the best interests of each Units training performance. In December of that year, the Department of the Army approved a syllabus drafted by the School for use on No. 1 Army Aerial Photographic Course. In February 1973 a 'Porter' aircraft was made available for the flying phase.

Navy photographers were not new to PHOTS. As far back as No. 30 Course, conducted before the formation of the Unit, the Navy had its 'finger in the pie'. Navy requirements saw personnel trained in photography for its 1947/48 Antarctic Expeditions, and for the formation of the new Fleet Air Arm. Photography in the Senior Service had previously been carried out by Department of Information personnel, any sailor with a camera, or friendly RAAF photographers. In March 1974, CPO PHOT Ron Batchelor, spent three days in discussions at PHOTS regarding the RAAF training Navy photographers. As a consequence, four members of the Royal Australian Navy Photographic Branch commenced No. 19 NCO Course on 2 October 1974. Some modifications to the syllabus were necessary, but caused no problems overall.

In May of 1975 the first of many Basic Photographic Courses was held which included Navy personnel. In fact, No. 65 Airmen's Course was the first course to contain students from all three Services (RAAF, NAVY, ARMY). No. 65 Course also had another first in that it started out as an 'Airmen's Course' and finished as a 'Basic Course' (owing to a change in course policy).

Although said to be impossible, by using a special mount made at RAAF East Sale, an F52 Aerial camera was fitted to the HS748 for use on the No. 18 Officers Course. A new era was entered during the year with the arrival of an Ilfospeed drier and the 'future death knell of the glazer was sounded'. Present students had to use the 'old fashioned' paper until stocks were depleted.

An Air Instrument Personnel Photographic Course of two weeks duration was held early in 1978, with instruction being mainly in aerial photography and the importance of photographic interpretation. The Comlab in Building 378 was completed at this time even though the eight De Vere enlargers did not arrive until May of that year. Included in the year's works program was the painting of a number of the School's buildings. The CO at the time expressed the view that their only hope was that 'they run out of pink paint before they started on any of the School's buildings.' The old Versamat building was a testament that 'they' did not!

Eleven members of the RAAF Requirements Committee visited the School in March 1978. While they appeared to show interest and supplied a sympathetic ear to the need for new premises, it was felt that it would be many years before the School could boast of adequate facilities. At one stage, the CO had a feeling that the plea for a new School had reached a higher, much higher, authority than could be expected. The month of June started badly when on the 3rd, Mother Nature decided to add to the frustration by blowing a tree through the School's one and only studio. This created no end of chaos until a former CFS classroom (Bld. 124) was converted into a studio. The floor in the classroom collapsed which necessitated complete restumping, but training continued with lessons being programmed at SAN. The new School was not forthcoming, but they did have a new and better studio and a much improved classroom.

In May 1978 the School's CO wrote in the Unit History: 'It was inevitable, but I admit a little sooner than I anticipated....today 25% - tomorrow 100%.' This comment was prompted by the first intake of female students. ACW Sue Robins, ACW Teena Snow and ACW Sue Flanagan received instruction on No. 70 Basic Course and became the first female photographers in the RAAF.

The First Female Photographers - ACW's Robins, Snow and Flanagan - March 1970

The posting in March of 1979 of an Administration Officer was received with 'jubilation and gratitude' - especially by the Chief Instructor, who really did not believe it until the new Admino was 'firmly nailed onto his allotted chair'. Later in August of that year, another course was instigated at the School with No. 1 Photographic Interpreters Familiarisation Course being placed into operation. The names were getting longer, but the courses were getting shorter. The course, of two weeks duration, enabled the graduate to become familiar with photographic requirements and terminology, thus providing him with the background to combine with photographic personnel on a technical level to provide imagery for interpretation purposes.

The Modern Era

For the first time, foreign nationals took part in a course held at the School when two Philippine Air Force Officers were trained on No. 22 Officers Course held between January and March 1980. Later in June of that year, preliminary work was completed on a new photographic magazine. The magazine, to be called 'Focal Point', was designed as a bi-annual publication to provide communication between members of the RAAF and RAN photographic world. To date, 'Focal Point' is still a going concern.

All through the early eighties, maintenance on the School was deferred more and more in belief that the new School was not too far in the future. Considering the original completion date for the School was to be late 1984, it is little wonder that any money spent on existing facilities would be wasted. Visits to large Service and civilian establishments were carried out to obtain further information for the proposed new School, however, it was becoming obvious that the new School completion date of late 1984 would not be met. Because of this delay, urgent maintenance was required on the existing buildings. Much of the design work involving the proposed new School is attributed to Flight Lieutenant Graham Walton, who was later to become the Commanding Officer of the new complex. Parliamentary approval for a new School was finally received in 1983.

Squadron Leader Peter Kelly assumed command in 1981, and it was during his reign that a project to gain unusual air to air photographs of the RAAF Aerobatic Display Team, the 'Roulettes', commenced. Discussions were held with MNTESL about the positioning of a Nikon F2 camera, fitted with a 28mm lens, motor drive and battery pack, at the rear of a Macchi aircraft. Graphics and optical calculations were given to CFS and planning was held on the positioning of the aircraft. It was decided that the best position for the camera was the speed brake door, this being dependant on the elimination of vibrations. Installation of the camera was completed in October 1982 and in April 1983 the system was used with outstanding success.

The first Photographic Resolution (Aerial) Course was held in December 1983. Staff at the School had expended enormous efforts in developing the syllabus for the two weeks training course. The practical side of the course was restricted through lack of equipment, and the absence of good aerial imagery, however an RF-111 from RAAF Amberley was made available to fly over targets laid by course members.

April 1985 saw the start of one of the largest courses to go through the School when No. 1/85 Basic Course began with a strength of 24 students. The course strength was to be a once only occurrence due to shortages of photographers in the service, but the trend continued into 1987/88. Due to course numbers, and the inadequate facilities at the School, no more advanced Courses could be held until the Unit moved into the new complex. The Unit got a shot in the arm during

May, with the arrival on strength of the first RAN instructor. The need for a Navy staff member had been recognised since the first attendance of sailors on the Basic Courses in 1975 even though higher authority had ignored all previous overtures for a Navy Instructor.

Squadron Leader Steve Clutterbuck assumed command of the School in January 1986 and one of his first observations at the start of No. 1/86 Basic Course, was that of the twelve direct entry trainees, eight were female. It seemed to those at the time that the mustering was, indeed an equal opportunity employer.

The great event that everybody had been waiting for occurred in October 1986, when construction began on the new School of Photography. Even though it started out as just a flattened site, it was heartening for the mustering to see such progress. The completion date of mid 1987 was misleading in retrospect, but at least a start had been made. Squadron Leader Walton assumed command of the School in July 1988. It was appropriate that the person responsible for the complex was in command while the complex was being constructed.

In December 1988 No. 1/88 Basic Photographic Course passed into history as the last course to graduate from the 1952 vintage School of Photography. Days after completion of the course, the Administrative side of the School moved into the new building, and into a new era of RAAF photography.

PHOTS Today

Today, the School of Photography continues the fine work of training for photographic excellence. The three Services and approved civilian students all benefit for the world class facilities and instructional staff.

The Modern School of Photography with the present CO, Squadron Leader Edwards discussing some of the finer points of the Sinar camera with Advanced Course students

CHAPTER 9

THE SCHOOL OF AIR TRAFFIC CONTROL

by Flying Officer Doug McLean

Military Air Traffic Control training commenced in Australia at RAAF Laverton during August 1946. The following year, this training was transferred to No. 1 Advanced Flying Training School at RAAF Point Cook, where it remained until late 1956. On 21 December of that year, responsibility for ATC training was again transferred, this time to RAAF East Sale, where it continued under the auspices of 'C' Flight, Central Flying School. From these humble beginnings, the need for an independent unit to train military Air Traffic Controllers became evident, culminating on 11 May 1981 with the formation of the School of Air Traffic Control (SATC) at RAAF East Sale.

Inauguration of the School of Air Traffic Control, 11 May 1981. Squadron Leader A.F. Chapman proudly leads the members of SATC towards their inauguration.

Establishing a 'New' School

The original request to establish an autonomous school was raised in 1979 by Wing Commander A.C. (Sandy) McPhie, when he was Command Air Traffic Control Officer at Headquarters Support Command, Melbourne. At the time of the School's formation, he was filling the senior ATC post in Air Force Office, Canberra.

'Of course, I was always confident the school would be established,' he said on being notified of the official date. 'There has been a significant increase in the training commitment since the RAAF commissioned five major ATC radar centres in the mid-1970's. But still, it is most satisfying to actually see the School come into being. I consider the upgrading in status to that of an autonomous unit to be well deserved recognition of a job well done and of the important role that air traffic control has in to-day's Air Force.'

The Commanding Officer-elect of the School, Squadron Leader Alan Chapman, was justifiably proud of the training record of Air Traffic Control students at the inception of SATC. With manning levels in the ATC category well below the infield establishment of 282 controllers, increased recruiting ensured the new school was fully utilised from the outset. The training commitment when the School opened for business in 1981, was for three or four ATC courses a year, each of 18 weeks and for 15 students, plus up to 12 radar controller courses of six weeks, with four students on each. Extra courses could be added as needed. Squadron Leader Chapman had nine instructors at the school. All were experienced controllers from the field. The instructors included Captain Graeme Lawler, an Army air traffic control officer, and Flight Lieutenant Ian Froome, who instructed in airfield surveillance radar techniques at the RAAF ATC training annex at Transport Australia's central training college in Melbourne.

The original buildings for SATC were converted World War II huts.

Staff levels increased slowly over the years until 1990/91 when under the guidance of the School's longest serving Commanding Officer, Wing Commander Ray McDonough, the establishment jumped overnight from 15 to 28 to accommodate increased courses brought about by a backlog of post-graduate training. The present staff of 25 includes, besides RAAF ATC personnel, education and training qualified computer specialists and administrative personnel and exchange instructors from the Australian Army, Royal Air Force and Royal Malaysian Air Force.

Training Aids

Many of the School's training aids have been made over the years by the instructors. School staff resorted to designing their own training aids to overcome the problem of giving air traffic controllers realistic training with established methods. This difficulty had faced the International Civil Aviation Organisation (ICAO) and member states for many years. Since 1949, the RAAF, like most other interested organisations, had used a number of aids and methods to teach students the techniques of controlling aircraft visually in an aerodrome traffic zone. These methods varied from model aircraft moving around on a system of motor driven pulleys and wire hooks, to electric trains (representing aircraft) operating on tracks suspended from the ceiling of a room, to animated films showing taxying, taking off and landing aircraft.

All suffered from the one basic handicap; at any time, only one student could receive any practice. Whilst one student controlled, others acted as 'pilots', operated projectors, or simply observed. This led to the design of what has arguably been the School's most successful 'invention', when in February 1964 School instructors, under the guidance of Squadron Leader Browne, Officer in Charge 'C' Flight CFS, designed an Air Traffic Control training system considered to be the first of its type in the world.

Known as the 'Audio-Visual Trainer', the system consisted of a room equipped with 12 cubicles, 13 headsets, 13 microphones, two tape recorders, recording tapes, a projector, slide synchroniser, projection screen, 35mm slides and a control console. Slides automatically flashed onto a screen, presenting the various situations as seen by a tower controller, such as aircraft at various critical circuit positions. Simultaneously, a master tape acting as 'aircraft pilots', requested the various instructions applicable to each slide. All students, hearing in their own headsets the pilot's request for taxying, take-off and so on, raised their microphones and answered the 'aircraft pilot' with the R/T phrase considered to be appropriate by each student. An instructor sitting at a control console monitored, corrected and recorded students' responses on a slave tape as considered necessary. At the conclusion of each sequence, the slave tape, synchronised with the slides, was played back to the students. A call light system enabled students to attract the instructor's attention should a question arise. In this event, the playback was stopped whilst the question was asked and dealt with by the instructor. Controls permitted all students to participate simultaneously during question time.

Recording equipment used to record and replay student's responses.

Maximum co-operation between East Sale units enabled the audio-visual trainer to be used on a limited scale from June 1964, until finally completed in December 1964. The RAAF School of Photography 'shot' 47 aircraft circuit scenes. From these, the ATC instructors prepared associated scripts. Realism was enhanced in the cubicles by displaying a photograph of a standard tower console requiring students to set altimeter and wind velocity dials and select a jet barrier to the appropriate position for a given situation. The greatest advantage of the system was that all students received training at the same time with little time wasted. The Audio-Visual Trainer, although updated over the years, is still in use today, and continues to provide very effective instruction in the phraseologies required in the Air Traffic Control environment.

A New Building

On the 16th and 17th of October 1987, staff and students relocated from the old SATC buildings which dated back to the war, to the new purpose built complex designed to house all School facilities with allowance for future expansion. Grouped around a central foyer and administration rooms, the building consists of three separate wings for staff, classrooms and simulators. Included in the building was a completely new surveillance and ground controlled approach radar simulator.

A major addition to simulation in the Air Traffic Control training environment, the RAAF ATC Radar Simulator, was commissioned on 19 October 1987 by Mrs Ros Kelly, MP, Minister for Defence Science and Personnel, at the same time as she opened the new building. This simulator was developed by Solartron Simulation UK and at the time was the most advanced training simulator of its type in the world. Solartron's technology provided unmatched design and operating flexibility together with a facility for almost unlimited reconfiguration or expansion. Still in use, the system is modular in both hardware and software. It can accommodate the full spectrum of modern ATC simulation requirements from a single pilot position and trainee radar display console with raw radar presentation up to an unlimited number of positions with fully synthetic display presentation.

The new School of Air Traffic Control building under construction.

The Minister for Defence Science and Personnel, The Rt Hon. Mrs Ros Kelly MP, opening the new School building on 19 October 1987, AVM Collings looking on.

The radar simulator has four 'Surveillance Radar' screens on which students learn the necessary skills to become radar approach controllers. Such skills include maintaining appropriate separation between aircraft while sequencing the aircraft to achieve an orderly and expeditious flow of arriving and departing traffic, maximising the use of available airspace while adhering to noise abatement procedures and keeping aircraft clear of hazardous weather or areas of known severe turbulence. The simulator also has three procedural consoles which are used to train students in the methods of controlling aircraft in areas outside radar coverage or during times of radar failure, and two precision approach radar screens which are used to provide talkdown for aircraft landing in poor weather conditions.

The computer generated targets (aircraft) on the screens are controlled by instructors and students who operate remote information and pilot stations. The instructor who operates the supervisor position has total control over each exercise to the extent of being able to add or delete aircraft as required; to fail radar frequencies, navigation aids or the radar itself; and to inject weather returns, radar jamming or emergency codes onto the radar displays. Each simulator console is equipped with similar facilities to those found in 'real life' control centres.

A Full Tower Simulator

There is full confidence that the School of Air Traffic Control's latest simulator, totally designed and manufactured in Australia, will stand the test of time. Commissioned by the Minister for Defence, Senator the Honourable Robert Ray on 19 June 1992, this latest acquisition by SATC is the Tower Visual Simulator (AUSTOWER).

Minister for Defence, Senator the Honourable Robert Ray commissioning AUSTOWER on 19 June 1992

The concept of AUSTOWER was developed by the Australian Government Aeronautical Research Laboratory and developed by Ferranti Computer Systems (Australia) Pty Limited. AUSTOWER is a full sized ATC tower environment with the external view, including moving vehicles, represented by computer graphics projected onto screens giving a view of 225 degrees horizontally and 34 degrees vertically. Realism is enhanced by computer controlled ambient lighting, vehicle sound, intricate model presenting, true vehicle spatial movement and a fully operational console with radar, airfield facility and communications controls.

The School uses this computer graphics aerodrome control tower simulator for training skills in Surface Movement and Tower Control. An exercise in software engineering, the Simulator provides realistic yet cost effective training. The system allows real time use of images with up to 10 vehicles visible at any one time. Students operating from a control cabin behind a large glass window at the rear of the simulator act as surrogate pilots and control vehicle movement by touch screen manipulation. The flexibility of operation is achieved by selecting movement choices from a database of options. Vehicles include fixed and rotary wing aircraft and such ground vehicles as fire trucks, ambulances and aircraft tow-motors.

Photo's were used to simulate the Tower environment. Students were required to use their imagination to visualise aircraft.

The teaching function of the simulator is enhanced by features such as record and replay, separation breakdown reporting, exercise preparation facilities and prompts for surrogate pilot radio-telephony and co-ordination purposes. Aircraft can disappear into a user set low cloud base or a reduced visibility. Night operations and transitions between day and night are available including vehicle and airfield lighting. Moving airfield artefacts such as arrester systems and windsocks are also a feature. SATC staff are able to manufacture new backgrounds, airfield environs, vehicle models and associated trajectory.

CO, Wing Commander Ian Batten, explaining the operation of AUSTOWER to Senator Ray.

AUSTOWER represents an improvement in ATC training - gone are the days of using photographs on partitions for a background and student imagination for traffic. The new device complements the SATC Radar Simulator, Flight Planning Office Simulator and the Audio-Visual Simulator for full ATC practical training.

The New AUSTOWER in operation

The Courses

The Basic ATC Course students at the School are drawn from all walks of civilian life and from serving military personnel. Men and women alike strive for appointment to commissions in the ATC category and after completing officer training at Point Cook, move to East Sale for the next scheduled ATC Course.

This course provides basic ATC training for the Australian Armed Services and for a number of overseas military personnel under the Australian Government's Defence Co-operation Program (DCP). The aim of the course is to produce a graduate trained in Aerodrome Control (Tower), Approach Co-ordination and Flight Planning Office. It is a very demanding course and requires total commitment by the students during the twelve weeks theory instruction in ATC and related subjects, and eleven weeks application of this theory in a simulated environment. On the job training in the 'live' ATC environment takes place on the base to which the student is posted on graduation.

Other courses conducted by SATC are the Advanced Air Traffic Control Course which provides intensive post-graduate training in both procedural and radar control. After completion, the graduates are qualified for employment as military approach/departures and en-route controllers and for employment in joint ATC centres located at capital city airports.

The Air Traffic Control Instructor Course provides specialist instructor training for members posted to SATC for an instructional tour. This course is also available for overseas military personnel and just recently, for controllers from the Civil Aviation Authority.

The Flight Planning Office Controller Basic Course has provided training for members of the RAAF Reserve in the duties of the Flight Planning Office Controller but now chiefly provides training for Service personnel from other categories and musterings who are selected for duty in the Flight Planning Office. The Air Traffic Control Short Course is designed for personnel with current air traffic control experience who enlist in the RAAF or for overseas exchange officers who require training in Australian military air traffic control procedures. The course is tailored to the needs of each individual student. The Ground Controlled Approach Controller Course which provides post-graduate training in final approach radar control, is gradually being phased out as ILS approach navigation aids are installed at RAAF Bases.

Two new courses are being conducted at SATC in 1993. One is an ATC refresher course for controllers who are returning to active controlling jobs after an absence of time. The other is the ATC Trainer course which will qualify controllers to act as training officers for their home bases.

Overseas Students

The cosmopolitan nature of the many courses is highlighted by the different nationalities that have been trained at SATC over the years. Graduates have come from the three services (RAAF, ARA and RAN), the civilian ATC community and from the respective Air Forces of Malaysia, Singapore, Indonesia, Nigeria, Pakistan, Philippines, Thailand and Brunei. Among the more distinguished graduates is a past Chief of the Air Staff, Air Marshal (retd.) Sir Neville McNamara, KBE, AO, AFC, AE, who completed No. 5 ATC Course in 1949.

Squadron Leader & Mrs Chapman greet AM Sir Neville McNamara AO,CBE,AFC,AE and Lady McNamara

Spirit of Competition

Although the 'old and bold' will always say how busy it was back in the 'good old days', the staff and students still have time for many sporting achievements. Over the years, members of SATC have competed in seven-a-side Rugby competitions, inter unit Cricket and Soccer matches, 'iron man' competitions, and 'boat races' to mention a few. More recently, staff have constructed a beach volleyball court and regularly hold lunch time volleyball matches and are presently constructing a half basketball court.

Members of SATC also support the local community sporting competitions with strong representation not only in the traditional ball sports, but also in ten pin bowling and volleyball. Additionally, SATC usually has a strong showing in the annual Marlay Point Overnight Yacht Race held over the Labour Day weekend. This race runs from Lake Wellington Yacht Club to Paynesville via the McLellan Straits and Metung.

The many members of overseas countries who pass through the school are taken under the wing of other students and instructors and are 'escorted' to places such as AFL Park in Melbourne to get a taste of Australian Rules football. Other leisure activities such as spending the day cruising the Gippsland Lakes have been highlights of their stay in Australia.

Welcoming a new member to 'Spooners' Volleyball Team at SATC's sand court.

Graduates

The first Air Traffic Control Course to graduate from the newly formed School of Air Traffic Control was No. 80 ATC course with eleven RAAF and two ARA students, in 1981. The most recent course was No. 113 when, on 18 September 1992, seven RAAF and one ARA students graduated. To date, the School has graduated 331 ab initio students and many more from post graduate courses. Many now serve overseas with Australia's northern neighbours. The School has established an international reputation as a centre of Air Traffic Control excellence, a reputation which is jealously guarded and will ensure a sound future for the Unit.

CHAPTER 10

BASE SQUADRON EAST SALE

by Flying Officer Robert Hayes

The Squadron Forms

Base Squadron East Sale was formed on the first of May 1952 from the Administrative Squadron of Headquarters RAAF Station East Sale as a separate Air Force Unit under direct command of the Officer Commanding RAAF East Sale. The function of the unit was then, as it remains today, to provide base services for all units located at East Sale. Services provided by Base Squadron include supply, catering, medical, dental, chaplaincy, welfare, physical training, transport, ground defence, fire, air traffic control, education and training, communications, electronic data processing and administration.

Wing Commander W.E. Townsend from Headquarters East Sale was appointed as the first Commanding Officer of Base Squadron in addition to his duties as Officer Commanding of the Base at the time. Wing Commander Townsend served as Commanding Officer until 20 February 1953, when he departed as a Group Captain on a posting to Eastern Area Headquarters.

The original strength of the unit on 31 May 1952 consisted of 22 officers and 205 airmen. As of January 1993 the unit strength was 37 officers and 264 airmen.

Wing Commander W. Townsend

A Country Estate?

There were many reports of hardships and overwork within the Squadron. By 1954 concern had been shown by a number of COs as to manning shortage problems within the Catering, Motor Transport and Equipment Sections. Many calls were made for civilians to fill these gaps, which were refused by higher authority. It is quite ironic that in 1991 plans were under way to cut the numbers of defence personnel employed in these same areas (along with other sections) and replace these positions with civilian equivalents.

Not only was there a problem with personnel numbers back in 1954. The Commanding Officer, Wing Commander Perrin was also concerned with the shortage and age of motor transport vehicles on the base. As he put it then, 'It is rather a curious sensation to fly from Sydney to Sale in about 40 minutes and then spend 20 minutes walking from the aircraft to the Mess because of the lack of serviceable transport'.

The Commanding Officer No.2 ACS - Group Captain Lings

Construction of the airfield was the major concern of Base Squadron in 1954. No. 2 Airfield Construction Squadron (2 ACS) relocated from Butterworth, Malaysia in August 1958 and began operations at East Sale in September of that year. Under the command of Group Captain P.G. Lings, No. 2 ACS remained at East Sale until May 1961. This unit was responsible primarily for the lengthening and improvement to East Sale runways and taxiways. When 2 ACS had completed their work, the runway and taxiway system was said to be the best in the RAAF.

Initial work by the squadron involved the excavation of gravel from river sites at Stratford and Boisdale and construction of storm water drains at the base. By early 1955 it was apparent that runway 09/27 had to be resurfaced, this task being completed in August of that year. Permanent lighting for the airfield was installed during 1956 and to complete the modernisation of the airfield, a concrete hardstand area was laid on the tarmac in 1957. This finally made the airfield suitable for use by jet aircraft.

On 30 April 1961 No. 2 ACS was officially disbanded, with a disbandment party and Detachment C of No. 5 ACS left behind to complete the remaining works and 'clean up'.

The Building Program

By 1957 building and works projects on the base were well under way. The Airmen's Mess and Base Canteen were completed and operational in June of that year. The development of Airmen's Mess facilities was a pet project of Wing Commander Alec Abicair MBE, Commanding Officer Maintenance Squadron at the time. Wing Commander Abicair was involved in improving the Base Cinema, constructing a children's playground and the establishment of the first 'wet canteen' for the airmen and airwomen of East Sale. So was born 'Abby's Tavern'. The club continued to develop into the A.J. Abicair Airmen's Club that exists today.

The Base in September 1958

Base Squadron suffered a major disaster in March 1958 when the Base Canteen was all but totally destroyed by fire. It took four fire crews, three from the Base and one from Sale, with a total of 50 men to eventually bring the blaze under control. An estimated £17,000 damage was sustained to the RAAF owned building and goods including electrical appliances and food all owned by the Australian Services Canteen Organisation (ASCO), the predecessor to today's Army and Air Force Canteen Service (AAFCANS).

There were many notable visitors to the base. August 1958 saw a visit to East Sale by the legendary world billiards champion Walter Lindrum. Lindrum gave Billiards and Snooker demonstrations in the base gymnasium in aid of the Anti-Cancer Fund, and was reported to have been an extremely popular visitor. When the swimming pool was opened in November 1960, Dawn Fraser the recent Olympic champion, provided swimming demonstrations for the benefit of several hundred spectators.

Work began on the construction of the Air Traffic Control Tower on 4 May 1959. Work on the Tower was progressing well until February 1960 when the contractor was declared bankrupt. Construction resumed with the use of 'day labour' and the Tower was completed in January 1961. AWA was contracted to install radios, this becoming a time consuming task. Air Traffic Control operations commenced from the new tower at 0800 hours on 18 April 1963.

Dawn Fraser is greeted by the AOC, AVM Alan and Mrs Walters and Wing Commander Fred Robey at the official opening of the pool in November 1960.

On 24 September 1961 the Minister for Air, Senator Harrie Wade, released details of money for new works at Victorian RAAF Bases. East Sale was allocated £131,000 for construction of the new Sergeant's and Officer's messes. July 1962 saw work begin. The Sergeant's Mess was complete and ready for use on 25 July 1963, the keys being ceremonially handed over to the Commanding Officer Base Squadron, Wing Commander Sullivan at a cocktail party that evening. The local press of the time reported the Sergeant's Mess to be 'the finest club in Gippsland'.

Construction of the Officer's Mess lagged behind that of the Sergeant's Mess. Wing Commander Sullivan observed that 'progress on the Officer's Mess appears to be at snails pace'. The extreme wintry weather was blamed for causing delays on most of the construction work being carried out on the Base. This work included the sealing and drainage of roads and paths and the building of an Airmen's Recreation Centre. The Officer's Mess was officially opened on 14 November 1963 by Air Officer Commanding, Air Vice Marshall C.D. (Doug) Candy CBE, co-incidently, the Base's first OC. The Airmen's Recreation Centre was finally completed on 10 September 1964.

The Cadet's Mess was completed in July 1969 and the sleeping and living quarters were occupied on the fourth of that month, even though the cadets were still eating at the Officer's Mess at this stage. By 1970 the dining room at the Cadet's Mess was operational, but by the middle of 1971 catering staff shortages caused the kitchen and dining room to close on a number of occasions. In March 1989 the Cadet's Mess became the Officers' Mess Annexe and today is an all ranks function area.

The Weather

Not only RAAF Units occupied the base. The Meteorological Section moved into their new building in February 1964. The office was officially opened on 28 February at a ceremony attended by the Officer Commanding, Group Captain N.W. Lampe, the Mayor of Sale and the President of the Avon Shire Council. A meteorological element has existed at East Sale since the formation of the base in 1943. Today the section is manned by full time civilians employed by the Bureau of Meteorology.

Speaking meteorologically, the harsh Victorian winters have always been a well documented 'bug-bear' of members of RAAF East Sale. This is quite understandable considering the large numbers of personnel originating from the warmer climates of Queensland, Western Australia and New South Wales. But anybody today who whinges about the weather whilst sitting in their temperature controlled work environments should spare a thought for their counterparts in 1965. Reports from June 1965 show that daily temperatures were down as low as 27 degrees Fahrenheit (-2 degrees Celsius), with many people working in buildings with limited heating from gas or oil burning fuel stoves. Electrical heating was not available due to the poor quality of electrical wiring in many of the wartime vintage buildings. It was not until December 1969 that work began on the installation of heating in the Airmen's and NCO's quarters, a great boost to morale at the time.

The Unit Badge

October 1973 saw the receipt of the Base Squadron East Sale Unit Badge after Royal assent by Her Majesty Queen Elizabeth II . The Base Squadron emblem carries the motto 'Support and Serve' and depicts a falcon's wing charged with a closed gauntlet. The gauntlet symbolises the unit as the focal point of all base activity and has a firm stronghold. The falcon's wing emphasises the unit's close association with, and support of, base flying activities.

One former member of Base Squadron who would have been well pleased with the choice of a falcon's wing as part of the Unit Badge was Harold Sebastian Wentworth Simm. One of the more noted, if not eccentric, personalities of the Squadron, Simm served at East Sale in the late sixties and early seventies before returning home to England in April 1973. He was reported to be interested in falconry, being a member of the Ancient Order of British Falconers.

Simm also claimed to be an Oxford graduate, having read English Literature in 1952, and to have been a commando in the British Army. He spoke with a strong Oxford accent and was apparently able to produce a photograph of himself at Oxford with several notable Australians including John Gorton and Gough Whitlam. As an LAC General Hand, Harold spent most of his time at East Sale as the permanent daytime guard on the front gate. He became a well known and popular figure to base personnel and to members of the Sale community in general, through his outgoing personality and antics such as taming the possums and birds that appeared around the front gate, constantly stirring up the Service Police and by jumping on the school bus for a sing-a-long with the children. But the most impressive episode of Harold Simm's colourful posting at East Sale would easily be the time he chased his boss around a yard with a shovel intent on doing him an injury. More guard duty!

Bravery and Crashes

On a more serious note, an exciting story of quick thinking and courage took place on 17 February 1978 when the actions of Base Squadron's Fire Controller, Sergeant Kevin Sloane, averted a major aircraft catastrophe. So impressive were his deeds that Sloane was awarded a national Bravery Medal. Sergeant Sloane had only been posted into East Sale's Fire Section for a week as the new fire and rescue controller when an incident occurred involving a RAAF Mirage aircraft. The Mirage was about to take off for a bombing exercise over the Dutson Bombing Range and was fully armed with four Mk 82, 500 lb bombs, as well as having 180 rounds of 30 mm ammunition and a full load of fuel.

Already being on stand-by due to the fact the aircraft was carrying ordnance, Sergeant Sloane was in a prime position in front of the fire station to notice that something was wrong as the Mirage made its take-off run. He ordered his crew into action without waiting for the crash alarm. The Mirage blew a tyre as it passed the fire station. By this time the pilot had aborted take off and the crash alarm had sounded, but Sloane and his crew in two fire vehicles, were already on their way in pursuit of the doomed jet.

When the aircraft hit the crash barrier net at the end of the runway, its nose wheel collapsed causing the fuel tank to split and burst into flames. After Flight Lieutenant Ian Watson had escaped from the aircraft, Sloane then ordered his crew in to put out the fire. Due to their immediate response the crew were able to smother the flames with a sheet of foam and extinguish the fire quickly, before it was able to detonate any of the jet's deadly cargo.

Following the incident, Sloane was full of praise for the work of his fellow crew members: LAC Gibbard Bucher, LAC Frank Conyon, LAC Adrian Thorpe and LAC Chris Wright. At an official ceremony the following year, Sloane was presented with the Australian Bravery Medal by the Governor of Victoria, Sir Henry Winneke. He became the first ever recipient of the medal, that had been introduced in 1975 under the Australian Awards system.

Another Base Squadron member was involved in a 'near miss' later that year, when on 13 December 1978 a CT4 Airtrainer crash landed on the base sports oval whilst delivering 'Santa Claus' alias Warrant Officer Tom Baker, as described in the CFS Chapter. Tom Baker, a regular when it came to playing Santa at the base and the last serving warrant officer pilot, donned the suit for 17 years. He decided to finally hang up his beard the year after his CT4 experience. Tom's retirement from the RAAF in 1979 was after 18 years service at East Sale, making him the longest serving person at the Base to date.

Santa's Aircraft 'after delivery'!

A Modern Base Develops

The late 1980's saw a great deal of construction of new buildings on the Base and the demolition of many old buildings, some dating back to the second world war. The first new building was Air Movements which was opened on 18 September 1985. During September 1986 Central Flying School Headquarters and the old officer accommodation buildings were demolished to allow for construction of a new CFS facility and the current School of Photography building. The School of Air Traffic Control also moved into their new building in June 1987 and Dental Flight entered their new premises in 1990.

A community employment program began on 18 May 1987 for the construction of an 'avenue of honour' and bicycle path along Aerodrome Road leading to the Base. Trees were planted along the edge of the road, each having a name plaque for Australian War Servicemen who were recruited from the Sale area. The Avenue of Honour was officially opened on 16 February 1988.

Air Traffic Control entered the modern age of high technology with the installation of the Selenia Surveillance Radar in 1988. The radar building on Deadmans Hill was completed in May 1988, while the initial installation of the consoles in the new approach control room were commenced. Final delivery of the radar antenna and its erection, began in September 1988. Final commissioning was in March 1989.

The last stage of works included a new Base Squadron Headquarters and Administrative complex which were officially opened by Air Officer Commanding Training Command, Air Commodore K.R. Blakers on 26 September 1991. The final building yet to be built will be Medical Section, completing the Base's redevelopment.

RAAF Base East Sale has much to boast about, largely due to Base Squadron efforts. A regular winner of the coveted `Tidy Towns' award, the Base now offers much to Service and Civilian employees and their families. Facilities such as an occasional child care centre, kindergarten and playgroup plus other standard amenities make East Sale a popular posting.

Today, Base Squadron East Sale continues to support and serve. 350 service and civilian employees are as proud of their role as their predecessors were, a role which remains vital for the smooth functioning of the establishment.

Base Squadron Personnel - December 1992.

CHAPTER 11

No. 32 SQUADRON

by Flying Officer Mark Diamond

The Original Squadron

In 1942, as the Japanese Forces pushed further south, the Hudsons of No. 32 Squadron become the eyes of the Australian defences and the aggressive sentries of the RAAF in the skies over New Guinea.

No. 32 Squadron was formed at Seven Mile Aerodrome in New Guinea on 21 February 1942. A month before, a composite Squadron comprising flights from 6, 23 and 24 Squadrons had been hastily thrown together to form the new unit. The original strength of 32 Squadron was 10 Hudson aircraft, 29 officers and 176 other ranks. Squadron Leader Deryck Kingwell was appointed the first Commanding Officer.

The Squadron began its operations immediately, with their first task being flown on the afternoon of February 21, the day of the Squadron formation. The first mission was an anti-submarine patrol in the vicinity of Murray Island. This was the first of many such flights against the Japanese in the Torres Strait and around the coasts of New Guinea and New Britain. On February 27, the Squadron engaged in its first bombing raid, a strike on Gasmata, a Japanese held aerodrome on New Britain. Five aircraft took part in the raid. Of the five, one returned with engine trouble, one was unable to locate the target and the remaining three aircraft successfully bombed the airfield.

The Japanese were resentful of the strikes being made upon them by allied aircraft, including the Hudsons and in retaliation began their own raids. Their objectives were Port Moresby and Seven Mile Aerodromes. As a result of this action, the Hudsons were dispersed as far afield as Horn Island. Seven Mile Airfield had now become a staging airfield for attacks on the Japanese in New Britain. On 7 March 1942, Flying Officer A.S. Hermes while returning from a reconnaissance mission over Gasmata and Rabaul, sighted a convoy of 11 ships. The convoy was heading for Salamaua. This sighting was the first definite indication of the increasing Japanese presence on mainland New Guinea.

Members of the original No. 32 Squadron at Horn Island, 1942

The Hudsons of 32 Squadron were the first to attack elements of this convoy in Salamaua Harbour. Squadron Leader Kingwell recorded a direct hit on an 8000 ton transport which was later seen to be burning and listing. During these raids, the Hudsons received their fair share of opposition from the enemy. Squadron Leader Kingwell himself evaded an enemy fighter over Salamaua on the 10th of March and evaded a further three over Gasmata on the 11th. However, two days later, he was not so lucky when he suffered slight damage from another five enemy fighters encountered over Port Moresby.

By 26 April 1942, the Squadron was based wholly at Horn Island. A detachment was also operating seaward from Townsville. On 6 May an important sighting was made by Pilot Officer P. Pennycuich in the Louisiades area when he reported seeing an Aircraft Carrier Group and later, four Destroyers and three transports. This was one of the first sightings connected with the battle of the Coral Sea. During August, a handful of Hudsons had been stationed at Milne Bay and were now involved in locating elements of the Japanese fleet. In September, the Squadron was withdrawn from New Guinea and by September 18th, was operating out of Camden, NSW. The Squadron was to be employed in anti-submarine patrols and convoy protection work. By the end of 1942, the Squadron had recorded 409 patrols and flown 615 flying hours. The squadron's strength at the end of 1942 was 47 officers, 320 other ranks and nine Hudson aircraft.

The opening months of 1943 saw 32 Squadron continue with the anti-submarine patrols. Enemy submarines were sighted as far south as Sydney. While attacks were carried out on the submarines that were sighted, there was no evidence of any destroyed. The Squadron also undertook searches for ships lost at sea. One of many was the search for the 'Lydia M. Child' which was torpedoed on the 27th of April. Although the area was covered extensively, no trace of the ship was found.

By the end of April, the Hudsons had been replaced by Beaufort aircraft and within a couple of weeks, there were detachments operating from Coffs Harbour and Bundaberg. On 18 June 1943, a Japanese submarine was attacked by Pilot Officer Harrison and his crew with a stick of depth charges. The Submarine surfaced in an apparently damaged condition and despite further attacks eventually disappeared. In May 1944 the Squadron's Headquarters was moved to Lowood in Queensland. The main task once again was the effective patrolling of the sea lanes stretching northward from Brisbane to Bundaberg.

1945 saw No. 32 Squadron carrying out security patrols twice daily in the Brisbane local area and the adjacent shipping lanes. Though considerably reduced in operational status, the Squadron was usefully employed for training. At war's end, the Squadron quietly disbanded at Lowood on the 30th of November 1945.

The Old and the New - Hudson and HS748.

The Squadron Reforms

On 1 July 1989 No. 32 Squadron was reformed at RAAF Base East Sale. Like the Squadron's original formation in 1942, the new 32 Squadron was formed out of two already existing flying units. Eight HS748s were transferred from the School of Air Navigation's old Flying Flight while a further two came from 34 Squadron in Canberra.

With the formation of 32 Squadron, the HS748s were removed from Training Command to became part of Air Lift Group under 86 Wing. As East Sale is a training base under the Command of the Air Officer Commanding Training Command, No. 32 Squadron effectively became a lodger unit.

The HS748's first prototype flew on the 24th of June 1960. In 1961 the HS748 series 2 prototype flew for the first time and it was in 1964 that the Australian Government announced that it would purchase the Series 2 for the RAAF. Eight Navigation Training aircraft (model 228) and two for VIP Duties (model 229) were ordered. The Navigation trainers were to replace the ageing DC-3 being used for Navigation training.

An HS748 on Navigation Exercise, Hobart, Tasmania

The first HS748 to arrive in Australia was A10-595. She touched down at RAAF Base Fairbairn on April 12 1967, where she remained until 1989. The first of the Navigation training aircraft arrived at RAAF Base East Sale on 20 September 1968 with the last arriving in August 1969. The Navigation Training aircraft were numbered A10-601 to A10-608 and the VIP aircraft were numbered A10-595 and A10-596. The two VIP aircraft remained as part of 34 Squadron at RAAF Base Fairbairn until they were both relocated to RAAF Base East Sale as part of 32 Squadron on 16 January 1990.

Wing Commander Peter J. Cleary, who had flown the HS748 as a Pilot Officer after completing his pilot training became the first Commanding Officer of the new 32 Squadron. The Squadron had a strength of 28 Officers and one Sergeant. In the first three years of the Squadron's operations, the strength has increased to 43 Officers and eight other ranks. Today the HS748s are not only used for Navigation training and VIP operations, but also for general air transport support to the Australian Defence Forces and for civil emergencies when they are so required.

With just under two months of operations, 32 Squadron was already being called upon to assist the civil community. On 25 August 1989 the Australian Federal Government invoked 'Operation Immune' in response to an Australian Federation of Airline Pilots industrial dispute grounding several thousand passengers. The HS748s were capable of carrying 26 passengers, plus two ground crew to act as 'Stewards'. The aircraft were employed on routes between Sydney, Canberra, Melbourne, Launceston and Hobart. Initially three HS748s operated out of Melbourne airport and this was later reduced to two aircraft. The aircraft were later moved to Canberra where they remained until the end of the dispute in December 1989. During the period of the dispute the HS748s of 32 Squadron carried over 25,000 passengers and flew over 1,000 sorties. The work was long and tiring, but wholly satisfying to those involved.

With the Squadron now being used more for passenger operations, the HS748s are beginning to spread their wings even more. The HS748s of 32 Squadron have been seen as far afield as Malaysia, Brunei, New Zealand, Port Villa and Papua New Guinea.

Members of No. 32 Squadron after 'Operation Immune'

Links with the Past

The links that are kept with the surviving members of the original 32 Squadron are very strong. The original CO of 32 Squadron Air Commodore Kingwell, now retired, has helped to keep the memory of the old 32 Squadron alive. In February 1990, Air Commodore Kingwell was taken to Port Moresby by the present CO, Wing Commander Peter Cleary, on the 48th anniversary of the Squadrons original formation, for the purpose of paying a tribute to those who paid the ultimate price. The Beaufort Association has also kept in close contact with the new 32 Squadron. Each year the association presents the Beaufort Association Trophy to a member of the Squadron. The recipient of the award is selected by the CO of the squadron and the presentation made by members of the Association on their annual visit to East Sale, where many of them learnt to fly during the war. The trophy is designed to recognise the consistent efforts and achievements made by individuals at the Squadron. In the past winners of the trophy have been:

Flying Officer Al Clements 1989
Flying Officer Bret Foxley-Conolly 1990
Flight Lieutenant Gary Westley 1991
Flight Lieutenant Dave Chaplin 1992

AIRCDRE Kingwell lays a wreath at the Port Moresby War Cemetery

On 7 August 1992 32 Squadron held a Squadron mixed Dining-in to officially present the Squadrons new Crest to a variety of guests. Air Commodore Kingwell flew down from Brisbane to receive a copy together with other members of the original 32 Squadron. Present were Messrs Col Butterworth, Bob Butler, John Murphy and Wes Watson. Several serving members present also received a copy of the crest. The crest depicts a Sulphur crested Cockatoo with its wings spread with the Southern Cross behind, all on a blue background. The design was developed by the members of the new 32 Squadron. Appropriately, the chosen motto is `Adaptable'.

Two founding COs - WgCdr Peter Cleary presents AIRCDRE Deryck Kingwell a copy of the Squadron badge.

Day to Day Operations

In April of 1990 the HS748 was again put to work in aid of the civil authorities. Flooding in and around Narromine, NSW, saw the HS748 used to carry 'change over crews' between Richmond and Narromine and also to carry spare parts for other aircraft on task in the affected area. The support that the HS748 provided meant that Hercules and Caribous were able to continue their flood relief without having to cease their operations.

September 1990 saw the eight Navigation training aircraft collectively accumulate 100,000 hours of flying without an accident. This milestone not only bears testament to the professional nature of the pilots who have operated these aircraft since 1968 but also to the dedication of the personnel at Maintenance Squadron East Sale.

No. 32 Squadron while avoiding major accidents, has not been without incident. During Operation Immune, Flying Officer Chris Colvin while landing at Canberra in heavy rain, hail and strong cross wind had the aircraft aquaplane and blew three of the main undercarriage tyres. He managed to pull the aircraft up on the runway without further trouble. Flying Officer Colvin was again in trouble at Mildura Airport after aborting a take off after a suspected bird strike on the take off roll. He unfortunately got bogged at the end of the runway when he tried to turn around on the dirt at the runway end. Flying Officer Colvin earned himself the name Christopher 'Crash' Colvin for his troubles.

Colvin was not the only one to have things go wrong. Flight Lieutenant Andrew Rae while enroute to Edinburgh, experienced an alternator failure which went undetected. This lead to a failure of the aircraft anti-ice system on one wing. The aircraft began to develop ice on the leading edge of the wing and around the air intake of the engine. The air intake became so congested with ice that the engine eventually 'flamed out'. Due to the increased drag and extra weight on the aircraft from the ice, they began to descend. As they descended the warmer air melted the ice and they were able to restart the failed engine. They diverted into Swan Hill with no further emergency.

Another victim of bad luck was Flying Officer Craig Phillis. While on his way to Norfolk Island in August 1992 to rescue another HS748 which had broken down, he happened to glance out of his window to find that the de-ice mats on his engine were on fire. The de-ice mats are electrically heated and are designed to melt ice which may form on the engine air intakes. The de-ice was turned off, the fire was extinguished and the aircraft was landed at Norfolk Island without further incident. Flying Officer Phillis was now himself in need of a rescue. 32 Squadron ended up having three HS748s at Norfolk Island of which two required rescues and one, the rescuing aircraft.

In January 1991, No 32 Squadron moved out of its temporary weatherboard accommodation, originally the airmen's mess into their new air-conditioned, brick Headquarters. The building was officially opened by Group Captain B.D. Harris, AM, Officer Commanding No 86 Wing, on 25 February 1991 and was previously used to house the School of Air Navigation Synthetic Navigation Trainer, now relocated within the SAN building.

On the 25th of March 1991 several members of 32 Squadron deployed to Horn Island, PNG for forward area training. The purpose of the trip was to try and locate the Camp that the original 32 Squadron operated from while resident in 1942. While the members were not successful, they did find aircraft wreckage, possibly one of the original Hudsons flown by the Squadron.

50th Anniversary Celebrations

On the weekend of 21-23 February 1992, 32 Squadron was involved in their own 50th birthday celebrations at Camden in NSW. For a short time, Camden was the Headquarters for 32 Squadron after returning from New Guinea. On the evening of the 21st of February members of the old and new 32 Squadron came together to enjoy an anniversary dinner, 50 years to the day of original formation. On Saturday the 22nd of February, the two groups again joined together for a march through the streets of Camden. At the end of the march there was a presentation of a Council Citation made to the Squadron and a Squadron Banner was presented to the Council. On Sunday, the final day of the anniversary celebrations, a service was held at the local Cenotaph to remember those members of 32 Squadron who were killed in the defence of Australia. Several wreaths were laid and a remembrance tree was planted in Macarthur Park. Naturally, during the service, there was a fly past of four HS748s.

A New Boss

On the 30th of March 1992, 32 Squadron changed command. While remaining part of Air Lift Group, 32 Squadron became part of No. 84 Wing, based at RAAF Base Townsville. Wing Commander Doug Graham the Officer Commanding No. 84 Wing, made his first visit to 32 Squadron on 16 May 1992 to familiarise himself with the Squadron's personnel and their operations.

On 1 July 1992 a new arrival to 32 Squadron signalled a large change to HS748 operations. Flight Sergeant Jim Zekants became the first ever Flight Engineer to be posted to 32 Squadron. He was followed two weeks later by Warrant Officer Dave Jones and Flight Sergeant Geoff Hill. These first three Flight Engineers spent their first six months at the Squadron doing a limited amount of flying and spending most of their time developing a conversion syllabus for Flight Engineers posted to the Squadron. By January 1993 seven Flight Engineers were on strength.

Another Aerobatic Team?

The public face of No. 32 Squadron is the formation team, the `Howlettes'. They are the only straight and level formation team in the world. Usually they perform as a four ship, but as does happen from time to time due to aircraft unserviceabilities they often appear as a two or three ship. The Howlettes do most of their work for Navigation and Air Traffic Control course graduations but have also flown at air shows and Anzac Day ceremonies. The formation team allows the pilots to use some of the skills learnt on pilots course which otherwise would not normally be used during their normal day to day work.

A Reputation Established

In three and a half years since reformation, the Unit has been recognised as an effective transport squadron which provides valuable support to the Australian Defence Forces.

The 'Howlettes' in Action over Gippsland

Squadron Group Photograph - 1993

CHAPTER 12

AEROBATICS TEAMS

by Flight Lieutenant Andrew Layton

Aerobatic pilots must be able to relax, and I might add concentrate intensely. The pre-plan sequence must not be altered.

- Squadron Leader Phil Dunn

It would be unfair to publish a book about RAAF East Sale without paying tribute to the RAAF's aerobatic teams. Until late 1961, Central Flying School (CFS) had never operated a permanent aerobatic team for flying demonstrations as there were few adequate staging airfields throughout Australia and among the suitable aerodromes, most were far apart. There were several aerobatic teams in the RAAF prior to the formation of the Red Sales in 1961 though most of those teams confined their displays to the south-east corner of Australia.

The Red Sales

The Red Sales was the first permanent aerobatic team established at CFS and it included four of the School's flying instructors. The Officer Commanding East Sale, Group Captain Peter Henderson, had recently returned from the United Kingdom and after marvelling at the precision aerobatic displays in England and France, was keen to establish a RAAF aerobatic team on the Base. Hence, the Red Sales evolved. The team derived its name from the red colour scheme on the nose and wing tips of the T-33 de Havilland Vampires and 'Sales' was added in recognition of the friendly relationship that existed between the Base and the City of Sale. Each pilot in the Red Sales underwent about a 30-hour training program and the team flew four T-33 de Havilland Vampires in formation for

GPCAPT Peter Henderson

their premiere public display at the East Sale open day on 19 November 1961. The team finished the year with displays at RAAF Base Laverton and Forest Hill (now RAAF Base Wagga Wagga) in December.

On 5 March 1962, the Chief of the Air Staff, Air Marshal Valston Hancock, outlined departm ental policy to all RAAF commands and headquarters: 'The following principles are to be rigidly followed when considering requests for participation of RAAF aircraft in flypasts or air displays: During Air Force Week each year, all requests for aircraft are to be refused except on those occasions of wide national importance or other special significance.'

The Red Sales started their 1962 display season on 14 June for the No. 22 Navigator Course graduation ceremony at East Sale. A Gippsland Times journalist who attended the parade described the display: 'The Red Sales flying wingtip to wingtip and trailing smoke, thrilled the audience of 400 with spectacular and brilliant aerobatics.' Flight Lieutenant Jim Rhind (RAF) was the first leader of the Red Sales and on 12 June, approved Flight Lieutenant Reg Jones to be his successor. The new team flew four practice displays during July and August in preparation for their public display at East Sale on the 16th of September. The fifth practice display on 15 August however, ended tragically when all four Vampires crashed during recovery from a formation barrel roll about 500 yards from Dutson Bombing and Gunnery Range, about 14 miles from Sale. The Vampires initially hit the ground about 50 yards on the south side of Dutson Road and the leading three aircraft hurtled across the road and disintegrated in the adjoining property. The fourth Vampire crashed about 600 yards away and all six aircraft occupants died instantly. Killed were Flight Lieutenants Peter Hearnden, Alec Young, Reg Jones, Marty Burke, Don Gow and Mike Dunbar.

In late October, a court of inquiry convened in Sale Coroner's Court to determine the cause of the accident. To assist the inquiry, the RAAF built a 3 foot mock-up wood and wire model of a jet aeroplane replicating a barrel roll. The court surmised that the 'flight path of the leading Vampire was so closely akin to that of the routine formation barrel roll until it had reached a dangerously low altitude, that No's 2, 3, and 4 did not realise that there was anything amiss until it was too late. No's 2, 3 and 4 did realise this in the very last moments and applied additional pull-up measures, but not soon enough.'

The Air Officer Commanding Support Command, Air Vice Marshal Douglas Candy, said from the evidence available, 'that the accident stemmed from a decision made by the leader shortly before or at the beginning of that different from normal manoeuvre. Whether or not the remainder of the team were aware of the change will never be known as the preflight briefing was normal and the frequency used by the Red Sales was not monitored.'

After the inquiry released its report on the accident, Hancock wrote to Candy: 'Formation aerobatic flying carried out by qualified personnel in suitable aircraft is a normal part of service flying training. Such flying practices pose a challenge and engenders in service pilots the qualities necessary to assist them in achieving complete control of their aircraft under exacting conditions.' Meanwhile, Group Captain Henderson wanted to resurrect an aerobatic team at East Sale but the Commanding Officer of CFS, Wing Commander Herb Plenty, preferred to 'let the boys (CFS staff) settle down for a while and let some of the memories lapse a little' before reforming a team in the School.

The Telstars

The Telstars became the CFS successor to the Red Sales and the original team formed officially on 27 February 1963 to present their debut public display the following day for No. 24 Basic Navigator Course at East Sale. Wing Commander Plenty said, 'the Telstars gave a highly polished display on their first public performance.' The team flew four de Havilland Vampire Mk35A jet trainers and their displays included a series of 'barrel rolls', 'loops', 'wing overs' and 'steep turns' at low-level with a 'bomb burst' to complete the sequence.

Telstars in Box formation

The Telstars conducted a flypast over Adelaide on 8 September 1966 as a preview to their display for the National Air Show at RAAF Base Edinburgh three days later. This interstate deployment was the furthest distance for any aerobatic team from East Sale. The display included 'loops' and 'barrel rolls' and their signature 'T for Telstar' formation. Squadron Leader Benny Raffin said that 'although rolling in "T" formation was difficult with the wingman flying line abreast of the leader, it provided a fine visual spectacle to the crowd below.'

In April 1967, the RAAF reviewed its commitment to aerobatic displays and on 27 June, the Air Board informed RAAF commands that display flying would continue but with the following changes: 'Annual displays at Laverton be rationalised to Laverton and Edinburgh on alternate years. Hobart Regatta retained on limited resources basis currently provided.'

On 29 June at East Sale, the Telstars trailed red, white, blue and bronze smoke from their silver Vampires during a display for the graduation parade of No. 33 Navigator Course. The use of smoke over the parade was a success but adverse weather prevented the team from flying their full display.

Between December 1967 to January 1968, the Telstars converted their aerobatic sequences to the new RAAF Aermacchi MB-326H aircraft which superseded the Vampires at CFS. The team's debut display in the Macchi was at East Sale for a School of Air Navigation graduation parade on 15 February and subsequent developments made it the final display for the team bearing the Telstar name. On 15 March 1968, the Chief of the Air Staff, Air Marshal Alistair Murdoch, recommended to the Minister for Defence that RAAF open days and flying displays scheduled for Air Force Week be withdrawn.

Following discussions between the Minister and the Air Board, AM Murdoch wrote to the RAAF commands on 19 April: 'The further need for the continued existence of display aerobatic teams as authorised by the Chief of the Air Staff is now suspended. To this end, the allocation of 495 hours requested by Headquarters Support Command for the command display aerobatic team will be withdrawn in the 1968/69 financial year. Formation aerobatic flying may still be carried out within the continuation training hours available. Aerobatic teams specifically for display flying are not to be prepared.' This decision was made in response to a reduction in flying hours allotted to the Macchi fleet after a slippage in the delivery of aircraft and a requirement to use the reduced hours for training at RAAF Base Pearce.

These developments led to the Telstars suspension on the 23rd of April 1968. Since the Telstars inception in February 1963, six different teams had presented a total of about 25 public displays. The last Telstars team conducted their final display on 18 May over Yallourn however, due to AM Murdoch's directive, the flypast was not publicised as a Telstars exhibition and official recognition for the team dissolved as these aerobats quietly passed into aviation history.

Almost two years would pass until CFS had an opportunity to form another aerobatic team and that chance came on 19 June 1970 when the Air Board discussed arrangements for the RAAF Golden Jubilee celebrations in 1971. The Board resolved that two aerobatic teams should be formed; Headquarters Support Command would provide a Macchi team and a Mirage team would belong to Headquarters Operational Command. Headquarters Support Command directed Headquarters East Sale to form a Macchi aerobatic team and authorised an additional 200 flying hours to meet the team's training and display requirements.

The Roulettes

Hence, CFS created the Roulettes in August 1970 which has become the most enduring team of its trio of aerobats. Flying the Macchi MB-326H, the Roulettes presented their premier public display in December for a RAAF Academy Graduation parade and in February, Squadron Leader Phil Dunn led the Roulettes to Tasmania for the Royal Hobart Regatta. Squadron Leader Dunn was an experienced aerobatic pilot to lead the team after flying with the Telstars and the Royal Air Force Red Arrows.

Four Macchis made up the original Roulettes

The Roulettes presented six displays by the end of April 1971 but their future beyond the Jubilee year was uncertain. In response to an inquiry from Headquarters Support Command regarding the Roulettes future, the Department of Air advised that the team would operate only for the Jubilee year. After a discussion between the Minister for Defence, Senator Tom Drake-Brockman, and the Chief of the Air Staff, Air Marshal Colin Hannah, Senator Drake-Brockman announced a change of policy on 20 September concerning aerobatic displays: 'Representative air displays and open days will be conducted at Pearce, Laverton, Edinburgh and Fairbairn during the weekend closest to 31 March 1972 and at Townsville, Amberley, Williamtown and Richmond over a similar period in March 1973 and that for the time being, we plan a similar pattern continuing in succeeding years.' This activity would: '...consist of demonstrations by individual or groups of representative aircraft and one Macchi aerobatic team from the RAAF CFS'. The original Roulette team demonstrated an eight-minute sequence of 'wingovers', 'loops' and 'barrel rolls' with formation changes such as 'box', 'tee', 'line astern', 'square' and 'swan'. Featured in the sequence was a synchronised 'pairs cross-over', or 'Roulette' manoeuvre which the team adopted as their name.

During 1972-73 the Roulettes performed mainly at RAAF bases across Australia with four Macchis and sometimes a fifth aeroplane accompanied the deployment between late 1973 to early 1974. The fifth team member mostly flew a solo display in addition to the four-aircraft formation. The display now included synchronised manoeuvres between a solo aircraft, two formation pairs and the main four-aircraft formation. The Roulettes premiered their five-aircraft display for the public at a RAAF Academy Queen's Colour parade at Point Cook in March 1974. The team then regularly performed the five-aircraft formations of 'card five', 'rhombus', 'kings cross', 'leaders benefit', 'vic', 'swan five' and 'line abreast'.

Roulettes now with smoke - an instant crowd pleaser

In September 1975, the people of Port Moresby, Kerema, Lae and Goroka viewed Roulette displays for the first time when the team deployed to Papua New Guinea for that nation's independence celebrations. For the trip overseas, East Sale contributed six Macchis and a Hawker Siddeley HS748 and personnel from CFS, Maintenance Squadron and the School of Air Navigation. The Macchis accumulated 150.6 hours during the detachment. Later that year, the Roulettes introduced a 'wine glass formation', 'upwards bomb burst' and an 'opposition loop' with a five-aircraft 'crossover' at the bottom. Then in 1976, the team added a 'thread the needle' manoeuvre that featured a 'head-on' pass between a solo Macchi and the other four-aircraft formation. The choreography then rarely changed until the next decade.

The Roulettes temporarily expanded to seven aircraft in 1981 to present a 24-manoeuvre, 16-minute display for the RAAF's Diamond Jubilee celebrations. The seven-aircraft formation presented the 'southern cross arrowhead', 'hexagon jubilee and donut' and solo and synchronised manoeuvres. This septuplet team performed in South Australia, Victoria and New South Wales prior to the premier jubilee event at RAAF Base Amberley in April. After that performance, the team reverted to the five-aircraft complement. Roulette Leader, Squadron Leader Brent Espeland, said the development from five to seven aircraft had doubled the complexity of the display which required unerring anticipation from team members to maintain a close formation. 'Because of the (Macchi's) lack of power, it requires enormous anticipation to keep a tight formation with seven aeroplanes. I only use 80 per cent of the available power during the seven-aircraft formation but the team had to move from idle power and speed brakes to full power and a prayer,' he said. The Commanding Officer, Wing Commander Ron Biddell, said the unit also did not possess the time, manpower or aeroplane resources to maintain a seven-aircraft team.

The Roulettes suffered their first accident on 15 December 1983 when Roulettes 2 and 3 collided at 2,000 feet in the team's training area near East Sale. Roulette 2 Flight Lieutenant Steve Carter, died instantly in the collision and Roulette 3 Flight Lieutenant Graeme 'Blue' Brooks, was separated from his Macchi as it disintegrated around him and he parachuted into a tea tree swamp on the 'Lakeview' property.

A sad day for the Base - OC, Group Captain Alan Pappin at a memorial service for the pilots, held outside the Officers' Mess.

An Esso helicopter medivaced Flight Lieutenant Brooks to the Base Medical Flight but during treatment he suffered a cardiac arrest and after cardiopulmonary resuscitation, he was transferred to the Gippsland Base Hospital where he later died. The training flight was the 25th of a planned 40 sorties and it was the second 'full show' practice for the 1984 Roulette team who planned to present their first public display at the Royal Hobart Regatta in February.

On 19 April 1985, five months after the accident, an altered 1984 Roulette team dedicated their first display at East Sale in memory of the deceased Roulettes, the late Flight Lieutenant Brooks and Flight Lieutenant Carter.

For Australia's Bicentennial year of 1988, the Roulettes planned to participate in about 25 events but a mid-air collision on 10 March between Roulettes 4 and 1 necessitated two unexpected changes to the team which postponed their public displays for several months. The Roulettes were practising for the Canberra Bicentennial show on 13 March and had allowed a Channel 10 news camera crew to film the sequences when the accident occurred.

As a result of this accident, one pilot ejected from his aircraft while the other pilot, Squadron Leader Jeff Trappett, managed to carry out a wheels-up landing back at the East Sale base. This action resulted in the Officer Commanding East Sale, Group Captain Doug Riding, praising Squadron Leader Trappett in the RAAF News for his skill in flying his badly damaged Macchi back to the airfield: 'He made all the right decisions getting back to the base, avoiding built-up areas and landing the aircraft as smoothly as if it was a routine landing.' Both pilots received minor spinal injuries in the accident and were admitted to the Gippsland Base Hospital for observation.

The Roulettes later formed a new team and resumed their bicentennial display program in August at Proserpine, North Queensland. In October, the team participated in the aviation spectacle of Australia's bicentenary with a five-aircraft display at the Bicentennial Air Show at RAAF Base Richmond.

During January to June 1989, a pair of Macchis presented displays in Victoria and Queensland as CFS staff converted to the Swiss designed Pilatus PC-9. The Roulette synchronised pair flew their maiden public display in April at the Phillip Island Motorcycle Grand Prix and in June, the Roulettes five-aircraft team presented their climactic display in Macchis over an appreciative audience at Lakes Entrance. In July, CFS started to form a Roulette team for the PC-9 but in the interim, the School conducted a series of synchronised pairs displays in the PC-9. A Roulette team training program between October and November led to a five-aircraft PC-9 display at East Sale in December.

During the ensuing months, Roulette Leader Squadron Leader Peter Barrett, sought approval to expand the team to six aircraft to partially compensate for the reduced noise impact of the turboprop PC-9 in comparison to the jet engined Macchi. The RAAF granted this request and the Roulettes six-aircraft display in the PC-9 made its debut at Mt Gambier in March 1990. To further enhance the spectacle of these displays in the PC 9, the RAAF arranged a repaint of the aeroplanes with a red and white design depicting a stylised red 'R' and the Southern Cross Constellation on the fin. The power and agility of the \$3-million PC-9 enabled the displays to be tighter than its Roulette predecessor but the Macchi was faster and more stable in turbulence and during formation flying.

The 1990 Roulette Team with a Roulette PC9-A featuring an early paint scheme. (L-R) SQNLDR Glen Steed, FLTLT Peter Clements, FLTLT Mark Scott, FLTLT Geoff Roberts, LEUT Mark Morey-Hype (RAN), FLTLT Tony Grady, SQNLDR Peter Barrett. SQNLDR Barrett was the first leader of the Roulette PC9 team.

For the first time in the Roulettes 21-year history, the team presented five aerobatic displays in the northern hemisphere at the sixth Asian Aerospace Exhibition, Conference and Airshow at the Changi International Exhibition and Convention Centre in Singapore between February and March 1992. The overseas deployment also featured the Roulettes first public displays with smoke in the PC-9. Several months before the Singapore deployment, the Chief of the Air Staff, Air Marshal Ray Funnell, tasked Hawker de Havilland to develop a system that would generate smoke for the Roulettes. The RAAF Aircraft Research and Development Unit tested a smoke pod system but chose to adopt an alternative method that would emit smoke by spraying a fine mist of oil into the aeroplane's right-hand exhaust outlet where it is heated and then condensed to produce visible smoke. The inclusion of smoke in the Roulette displays have since brought a new dimension to the show and has continued the tradition of aerobatics with smoke presented by the Red Sales, Telstars and earlier Roulette teams who flew the Macchi.

CFS has now provided formation aerobatic teams that have spanned four decades. The Red Sales pioneered aerobatics at the School and then the Telstars refined the displays which have continued to evolve with the Roulettes. The era of the Roulettes has generated wide exposure for the RAAF to the public throughout the world and each display demonstrates the high level of flying skills that a military pilot can attain.

Roulettes taxy for another show

CHAPTER 13

THE STORY CONTINUES - 1993 to 1999

by Wing Commander Keith Miller

Since the first edition of this book and the Open Day on the 18th of April 1993 to celebrate the 50th Anniversary of the Base, the support elements of the Base have undergone enormous change, while the four schools have endured essentially untouched, but with some changes to roles and tasking. Base Squadron East Sale and Maintenance Squadron East Sale no longer exist, the Base has a new unit called Training Aircraft Logistics Management Squadron, and most Base support services are now provided by civilian contractors or Australian Public Service staff.

These extensive changes to Base support resulted from a RAAF-wide review of Air Base support in the early 90s, and later the Defence Efficiency Review, the Commercial Support Program and the Defence Reform Program. The thrust of these policy initiatives was that Defence budget savings would be made from civilianising or contracting out all support tasks that did not require uniformed personnel to perform them. The thrust was also to focus the reduced numbers of uniformed personnel towards direct combat-related tasks.

This chapter will describe the far-reaching changes to East Sale Base support arrangements, and provide an update on changes to and highlights of the training schools over the past six years.

The Demise of Base Squadron East Sale

The demise of Base Squadron East Sale was brought about by the review of Air Base support which essentially recommended that base headquarters and base squadrons be combined to form air base wings. When Wing Commander Bob Harris arrived at Headquarters East Sale in January 1993, little did he realise that he would be the last Administrative Staff Officer. The post was disestablished at the end of 1993 and Wing Commander Harris then was posted to be Commanding Officer Base Squadron East Sale in January 1994. In July 1994 Base Squadron East Sale became Number 305 Air Base Wing, commanded by the then Base Commander, Group Captain Jeff Day, with Wing Commander Harris as Base Executive Officer.

The End of the Line - Base Squadron East Sale Commanding Officers Board.

Therefore, Wing Commander Harris had the unusual distinction of being the last Administrative Staff Officer, the last Commanding Officer of Base Squadron East Sale and the first Base Executive Officer, all in the space of 18 months.

The name change to 305 Air Base Wing had little impact on base support; the major changes were just around the corner, and will be described later.

The Closure of Maintenance Squadron East Sale

The late 80's had already introduced contractor support to the Base with Hawker de Havilland providing maintenance support for the PC9/A aircraft. Although this contract was short-lived, the seeds had been sown and in April 1994 Airflite Pty Ltd commenced a long term contract to provide PC9 maintenance support. When combined with the departure of the CT4 aircraft from the Base at the end of 1991, this meant that command and administrative support infrastructure and personnel in Maintenance Squadron East Sale were now in place to support the maintenance of only one aircraft type (the HS748), plus ground support equipment maintenance.

Sadly, particularly for the many people with such fond memories of their service with the unit, the decision was taken to close Maintenance Squadron East Sale and transfer the HS748 maintenance task to 32 Squadron and the ground support equipment maintenance task to 305 Air Base Wing. Maintenance Squadron East Sale closed on the 14th of July 1995. The Governor-General's Banner is now displayed in a cabinet in the Sergeants' Mess, an appropriate place in consideration of the many former unit personnel who live in the local area and visit the Mess.

Maintenance Squadron East Sale - The Last Unit Photograph.

Training Aircraft Logistic Squadron (TALMSQN) is responsible for the delivery of Integrated Logistics Support (ILS) to the ADF's HS748 and PC9/A fleets. TALMSQN's PC9/A customers are Central Flying School (RAAF East Sale), No 2 Flying Training School (RAAF Pearce), Aircraft Research and Development Unit (RAAF Edinburgh) and No 77 Squadron (RAAF Williamtown). HS748 customers are based at 32 Squadron (RAAF East Sale) and 723 Squadron (Naval Air Station Nowra).

TALMSQN was formed as a result of the 1992 Headquarters Logistics Command (HQLC) reorganisation which created the Weapon System Logistic Management Squadron (WSLM) concept of collocating the ILS provider with the principle weapon system users. Elements of Support Group 4, Aircraft Engineering and Aircraft Equipment Engineering Sections, and Staff Officer Repair and Overhaul were combined in August 1992 to form TALMSQN under the management of Mr Terry Carmody.

The Squadron relocated from Melbourne to East Sale in December 1993 with WGCDR Jim Hurley in command. In its formative years TALMSQN concentrated on developing its systems and procedures to meet Air Force directives. TALMSQN quickly became recognised as an innovator as the Squadron moved to the forefront in the development, trialing and implementation of Weapon System Databases and Engineering Databases. These efforts were acknowledged with TALMSQN being awarded the Markowski Cup for 1994 and a Defence Quality Award for 1995.

WGCDR Gary Marks was posted into TALMSQN as CO in January 1996 and under his command TALMSQN continued to chalk up firsts being the first WSLM to achieve Authorised Engineering Organisation status in March 1996 and AS/NZS ISO9001:1994 Quality Management System accreditation in May 1996. Additionally TALMSQN again was the first WSLM to negotiate Design Authority Contracts with its principle Weapon System Original Equipment Manufacturers, British Aerospace Regional Airways in June 1996 for HS748, Pilatus in November 1996 for PC9/A, and Rolls Royce in April 1997 for HS748.

In June 1997 TALMSQN took on the added responsibility of being contract manager for the leased B-200 (King Air) Aircraft that support School of Air Navigation training. This contract management role continues to grow with management responsibilities for additional leased aircraft based in the Northern Territory. Additionally, TALMSQN has been given the responsibility to transfer all leased aircraft from the civil to state register.

TALMSQN became an element of the Support Command Australia (SCA) organisation on the 1st of July 1997 and in October 1998 the East Sale Mechanical Engineering and Operational Maintenance and Ground Electronic Services sections were transferred to SCA and were placed under the Command and Control of CO TALMSQN. With the HS748 to be phased out of service around the end of 2000, the face of TALMSQN will continue to change.

Contractor Support

The introduction of contractor support at East Sale commenced with Ferranti support for training simulator maintenance in the 80's, and continued with Hawker De Havilland and Airflite Pty Ltd for PC9 maintenance and Lloyd Helicopters for search and rescue support. However, it was in the mid 90s under the Defence Commercial Support Program that commercialisation was given full rein. The guard roster became history in June 1994 with the letting of a contract for Base access control to Dale's Security, but in June 1996 the biggest change to support arrangements in the history of the Base was introduced.

SERCo-Gardner Merchant were awarded a support services contract which covered the provision of catering management, unit store and warehousing management, mess administration, cleaning, pest control, waste disposal, grounds maintenance, asset management, aircraft refueling, airfield sweeping and air terminal services. The only uniformed catering person at East Sale now is the Warrant Officer in the contract management cell. The cultural change was enormous, but three years into the contract the new arrangements are working well, and the teamwork and goodwill that always allowed things to happen at East Sale with a minimum of fuss is still evident.

Another contract was awarded to Hadens FM in March 1997 for fixed plant and equipment maintenance. P & O Asset Services hold a contract for general building and facilities maintenance, which was previously the responsibility of the Commonwealth through Asset Services, which no longer exists. Since June 1994, Honeywell Ltd has had the contract for simulator maintenance at SATC and SAN.

The level of contractor support will increase with the HS748 replacement to be contractor provided and maintained, and a contract covering fire services, access control and transport fleet management expected to be let late this year. Additionally, contractor provision of clerical and administrative services for the Base is also in the offing. Perhaps a contractor will write the next upgrade to this book!

SERCo-GM Staff at Work

Defence Reform Program

The next major cultural change to East Sale was the implementation of the Defence Reform Program. From the 1st of July 1997, the Defence Department was divided into 14 programs or major functions and each program became responsible for all like assets throughout the Defence Force. Accordingly, it came to pass that the Head of Defence Corporate Support Southern Victoria became responsible for the provision of most base support services at East Sale, including those already contracted out.

For uniformed personnel working in those elements taken over by Defence Corporate Support it meant that one day they were working for the Base Commander and the next they worked for a civilian organisation based in Melbourne. Not surprisingly, the past few years have brought some confusion and change fatigue for many East Sale personnel.

Other changes include the Defence Estate Organisation taking over responsibility for asset management, Defence Information Group has taken over responsibility for communications and information systems and Support Command Australia has taken over responsibility for telecommunications and ground support equipment maintenance and engineering support.

These latest changes meant that the Base Commander and Executive Officer were left with few uniformed support personnel still directly responsible to them, but with coordination responsibilities for various service provider programs. However, the structure of Base management has also changed, as discussed later.

In the longer term these changes will see the uniformed personnel replaced by APS staff or contractor staff or by a telephone or computer screen to access information and services centralised in Melbourne or elsewhere. For instance, married quarters are now allocated from Melbourne and household removals on posting are now arranged by personnel direct with Removals Australia over the telephone.

Central Flying School

Central Flying School has managed the usual heavy instructional, examining and Roulette workload over the past six years without accidents or major incidents, but there have been changes and highlights. Some of the highlights for CFS include being awarded the Markowski Cup in April 1997, a major efficiency award, achieving all targets in 1996 despite the loss of half the unit's qualified flying instructors to civilian flying jobs, and the celebrations of 25 years of the Roulettes in 1996 and the 85th anniversary of Central Flying School in November 1998.

The major change concerns the return of the CT4B aircraft to CFS. The Base last saw CT4B aircraft in December 1991 when the fleet flew to Point Cook, seemingly for the last time. However, two CT4B aircraft flew into East Sale late in June 1998, to enable instructor training to begin under the new basic flying training arrangements at contractor British Aerospace Flight Training Australia premises at Tamworth . Three more CT4Bs arrived at East Sale early in 1999.

Another important change for CFS with the new arrangements for ab initio flying training is that the unit is now responsible for the oversight of all flying instruction in the Australian Defence Force.

School of Air Navigation

In 1993 the school took on the additional task of training of intelligence personnel and also airman aircrew. While the Intelligence Officer and Intelligence Analysis courses only lasted one course each before being transferred to the army, the Airman Aircrew Basic course continued at the rate of three to four courses per year until the end of 1997.

In 1993 the school was approached by the Royal Malaysian Air Force to conduct a special navigator course to train aircrews for the newly purchased B200 King Air Maritime Patrol aircraft. Due to a tight schedule the course had to be completed in 31 weeks. As the students were going to be operating in a maritime environment, the training could be specialised to the maritime environment and the course length could be reduced without a significant reduction in graduation standard. With the help of 3 RMAF instructors a graduation rate of 70% was achieved, which was a significant improvement over the historical success rate for foreign students.

Also new to SAN in 1993 was the commencement of New Zealand Navigator and Air Electronic Operator training. Both of these aircrew categories were formerly trained at RNZAF Base Wigram near Christchurch. With the closure of that base the decision was made to combine RAAF and RNZAF navigator training at SAN with the AEOPs utilising the HS748 for the airborne part of their course. As part of the training exchange the RNZAF also supplied Navigator and AEOP instructors, HS748 pilots and clerical support.

The role of the navigator was rapidly changing, and the first significant recognition of that change came in 1995 with the decision to discontinue astro-navigation training as part of the basic navigator course. There were mixed feelings with the phasing out of the instruction in astro-navigation. The development of the satellite navigation systems over a number of years had provided ample warning of this event. This one change, however, marked the end of a significant link to former navigators that could be traced back many centuries. While the use of satellites provides significant benefits in all aspects of air navigation, the self satisfaction of successfully fixing the aircraft's position through use of such a basic technique will now be the preserve of navigators past.

During the next couple of years the school experienced increasing serviceability problems with the HS748. This in turn delayed graduation dates and even resulted in the cancellation of some courses. With the outlook for navigator and observer numbers in the ADF set to increase, a solution to increase the training throughput was needed. The idea of a replacement for the HS748 was not at that stage an option. In 1997 a decision was made to lease two modified B200 Super King Air aircraft to supplement the HS748 in the navigator/observer training role. An additional two B200 Super King Airs will be added to the fleet in May 1999.

B200 Aircraft

The role of the new aircraft was to provide an introduction to the airborne environment by having the student occupy the copilot seat. This aircraft also proved to be a superior platform for the instruction of overland tactical flying so it quickly adopted the tactical training role. Initially the aircraft and pilots were part of the school which meant that SAN had regained in some part its former 'flying flight'. This situation was to be short lived with the aircraft and pilots being transferred to 32SQN in 1998.

The decision was finally made in mid 1998 to replace the HS748 with a new navigation trainer aircraft. In March 1999 the Air Force decided to replace the HS748 with a fleet of four Beech 1900D aircraft and two Beech B200 or 350 Super King Air aircraft. The first of the new aircraft, to be operated by 32SQN could be ready as early as mid 2000.

In preparation for the new aircraft, but also to update navigator training generally, the basic course has been rewritten to align the flying exercises more closely with the operational skills required of the modern navigator. Gone are the long, high level, straight line plotting exercises so much a part of the basic course for many years, and in are shorter more intense mission orientated exercises that emphasise systems management and airmanship tasks. The Navy Observer course also is changing significantly with the flying exercises now to be conducted from the right hand seat of the B200, reflecting the tasks and skills required of future Navy Observers operating in naval helicopters. Both of these changes are being introduced during 1999.

School Of Air Traffic Control

The last six years have seen many changes for SATC, but 1997 was arguably the year in which the most dramatic changes occurred. On the 14th of April, four civilian students from Airservices Australia (AsA) commenced training on number 123 ATC Course. This marked the start of joint military/civil ab initio tower training. Staffing at SATC reflected this shift in focus with four AsA instructors based at East Sale. This training continued until July 1998 when AsA decided to cease ab initio training of tower controllers. The last AsA students graduated with 126ATC on the 9th of July 1998. AsAís decision to cease training was not because they were disatisfied with SATC, but because they had satisfied their training needs, and by a decision to examine the feasibility of privatising some of their Tower Control operations thereby reducing their controller requirements.

Another historic event in 1997 was the restructure of the provision of military ATC in Australia resulting in the RAAF assuming responsibility for all military ATC services. This brought to an end 35 years of Army ATC in Australia with the last Army students graduating with 122ATC on the 26th of June 1997.

With the RAAF becoming more focused toward combat operations, SATC altered the focus of basic ATC training to provide a more combat ready product than had been produced in the past. Tactical ATC training was introduced for 125 ATC in January 1998 coinciding with the amalgamation of the various Base ATC Flights to the command of 41 Wing. This phase of training was designed to introduce students to battlefield ATC operations and provide an introduction to how they could expect to be employed in a combat environment. The tactical training provided at SATC also provides the basic framework for further training by 41 Wing.

Late in 1998, construction was commenced on a new building to house a radar simulator which will meet the requirements of Project Air 5186, the Australian Defence Air Traffic System (ADATS). Installation of the new simulator is expected to commence in the year 2000.

Since the first edition of this book, SATC has conducted 15 Basic ATC courses and graduated 165 ab initio controllers.

SATC is currently commanded by WGCDR Peter Watson and has a staff of 27, which includes exchange officers from Malaysia and the United Kingdom.

School of Photography

The early 90's saw the School of Photography continue with the work of training for photographic excellence, under the command of Squadron Leader George Edwards. Throughout this era, the School began to incorporate new technologies electronic imaging and slide-tape presentations to enhance the services provided to the RAAF and meet customer requirements. This was done by focusing on the application and use of audio/video visual technologies. These were to have far reaching effects on the Defence Force as a whole. Instruction in audio visual applications were incorporated into the Advanced Photographic Course.

In January 1996 Squadron Leader Ron Vansetten assumed command of the School. During his command the unit entered the commercial support market testing program, which was a challenge for the School to face. In preparation for market testing the unit rewrote and reorganised course syllabi, gathered data on unit activities and helped prepare the statement of requirement. Nevertheless, the unit continued to rewrite and reorganise course syllabi and train students from all three services and from overseas, as was the case on the 1/97 Aerial Resolution Course, with one student from the Royal Malaysian Air Force.

During this period the RAAF gave recognition to the importance of digital imagery and thus after writing a new syllabus the School conducted its first digital imaging course in early 1997. The introduction of digital imagery to the RAAF was, as one Sergeant Photographer stated, 'like the introduction of colour television to the world', as it placed the Air Force on the cutting edge of technology and allowed almost real time photo acquisition.

Squadron Leader Greg Mathie assumed command on the 12th of January 1999 and also in January 1999, School of Photography became directly responsible to the OC Ground Training Wing, following the restructure of Training Command units.

Though providing photographic excellence remains the School's number one priority, the unit still remains under the commercial support market testing program with a final resolution expected by the middle of 1999.

Number 32 Squadron

A major highlight for 32SQN in the past six years was the Governor General of Australia awarding the following Battle Honours to the squadron on the 13th of March 1994: New Guinea 1942; Moresby 1942; Milne Bay; Pacific 1942; Rabual 1942 and Coral Sea.

In July 1995, with the closure of Maintenance Squadron East Sale, 32SQN assumed the responsibility and the resources for the HS748 maintenance task, and also found a new and permanent home in the previous MNTESL headquarters building.

The role of the unit changed in 1997, largely due to the increasing cost of maintaining the HS748 aircraft, whereby the squadronís responsibility for general transport tasks was removed, and SAN training became the sole flying task of 32SQN. The squadron continued to support ADF deployments to Papua New Guinea and the Middle East by sending personnel.

In 1998 the two Beechcraft B200 Super King Air aircraft operated by SAN were transfered to 32SQN so that all aircraft operations in support of SAN could be consolidated into a single squadron. Later in the year 32SQN also assumed responsibility for the Twin Otter transport support operations based in RAAF Tindal. This aircraft is expected to be replaced by another Super King Air during 1999 when the unitís fleet of these aircraft expands to five.

A major change for 32SQN in 2000/2001 will be the withdrawal of the HS748 from service and its replacement with a contractor maintained aircraft, the Beech 1900D

Changes to the Higher Base Structure - the Air Training Wing

From January 1999 East Sale became the home of the newly formed Air Training Wing of the RAAF, which encompasses all air training units in the RAAF. These include the three air training units at East Sale, as well as 2FTS RAAF Pearce, the ADF Basic Flying Training Unit Tamworth, the RAAF Institute of Aviation Medicine RAAF Edinburgh and the Combat Survival Training School RAAF Townsville.

The new entity is commanded by a Group Captain post, formerly the RAAF East Sale Base Commander post. Group Captain Paul Moran was therefore Base Commander one day and then OC of the Air Training Wing the next, with Australia-wide responsibilities. The OC has a Headquarters staff of seven to help him manage the Air Training Wing.

Another new entity, or an old one with a new name, was also formed in January 1999 when Combat Support Force East Sale (CSFESL) replaced 305 Air Base Wing. CSF (not to be confused with another East Sale unit CFS) comprises the remaining uniformed elements of 305 Air Base Wing and is commanded by the OC CSF who becomes the Base Commander. The OC CSF post was formerly the Base Executive Officer post, and the first man to fill the Base Commander post at Wing Commander rank, on a substantive basis, since the early 50's was a local lad, Maffra born and raised WGCDR Alex Danischewski.

The RAAF Balloon Provided Tethered Flights for Spectators at the 75th Anniversary Open Day.

Other Changes and Highlights

A significant change affecting East Sale air traffic controllers was the formation of Number 41 Wing at RAAF Williamtown early in 1997. The provision of air traffic control services at East Sale, and all other flying bases in the Defence Force is now the responsibility of 41 Wing, through 41 Wing Air Traffic Control Flights. This is yet another example of a base support service whose provision is now the responsibility of an off-base agency.

Major highlights for the base were the two open days, on the 18th of April 1993 to celebrate the 50th anniversary of RAAF East Sale, and on the 24th of March 1996 to celebrate the 75th anniversary of the formation of the RAAF. The weather was perfect on both days and each attracted crowds of 12-15,000 people.

Many more of the World War 2 buildings were demolished over the past six years, including the Australian Rules Football clubrooms, but the facilities highlight was the construction and opening of a new Medical Flight building on the corner of Battle Avenue and Catalina Street. The building was officially opened on the 12th of October 1994.

New Medical Flight Facilities Opened in October 1994

Sadly, because of the rapidly decreasing numbers of uniformed personnel at East Sale, the Base no longer fields Australian Rules or cricket teams in local competitions. However, the Sale Rugby League Club, which had the Base as home ground and has RAAF members as the major component, has been very successful in the local competition, winning premierships in 1995 and 1997. However, this club has now also ceased to be a RAAF Welfare sponsored club and moved to an off-base home.

The Base Emergency Force, although much smaller in numbers than in earlier years, has been very active in assisting local authorities during civilian emergencies over the past few years. The force has assisted with flood relief at Paynesville, been on standby for firefighting support and assisted the civilian police in the aftermath of the Longford gas disaster. The team also has a vital role to play in the now annual Base security exercises and in SAR exercise, as well as providing support for Officer Training School exercises which are now held regularly in an area near the range.

The relationship of the Base to the local community is as strong as ever and has been reinforced by the number of contractor employees from the local area now working at East Sale. This trend will continue. The good relationship with the community was demonstrated ceremonially on the 2nd of October 1998 when the Base exercised its Freedom of Wellington Shire, marching along Raymond St Sale, accompanied by members of the 4th/19th Prince of Wales Light Horse Regiment. The highlight of the day was a magnificent flypast by all 16 PC9 aircraft at East Sale, a tribute to CFS and to the good management of contractor Airflite.

The spirit and goodwill that has always enabled things to happen at East Sale with a minimum of fuss still abounds, however even the best laid plans can go astray and a touch of humour is never far away. Such as when SerCo-GM grounds supervisor Mick Symons caused great consternation amongst the catering staff when his grounds staff commenced preparations for a SAN graduation a week early. Or when golf club stalwart CPL Jason Campbell discovered snakes on several greens when he was watering late one afternoon. He alerted the RAAF Police of the invasion and commenced attacking one of the reptiles with a stick only to find out that the realistic-looking snakes were plastic, and had been provided by another club member to keep birds off the greens.

The Future

The future of RAAF Base East Sale seems assured, despite rumours from time to time in this current climate of sweeping changes in the ADF. In the short term, the numbers of uniformed personnel performing support functions will continue to diminish, but the potential for other elements of the RAAF, and particularly from Training Command, to relocate to East Sale is high and has been under discussion for some time.

The two major changes on the horizon are the possible commercialisation of the training of photographers for the ADF and the replacement of the HS748. A decision on the future of the School of Photography will be known soon as will whether or not a contractor awarded the contract will want to use the existing School of Photography state-of-the-art facilities.

The uniformed population will decrease dramatically once the HS748 is phased out of service, around the end of 2000, as about 100 RAAF personnel are involved in HS748 maintenance. The replacement aircraft will be maintained by a contractor.

Therefore, despite the many changes and the uncertainty on School of Photography's future, RAAF Base East Sale continues its vital role of training aircrew and air traffic controllers for the ADF, and will do so well into the next millennium. The Base continues to support exercises, such as the navy Fleet Concentration Period and bombing at Dutson AWR, and remains a vital asset in the projection of air power in the defence of Australia.

ANNEX A

OFFICERS COMMANDING RAAF EAST SALE

GPCAPT C.D. CANDY	APR 43 - JUN 43	
WGCDR K.S. HENNOCK	JUN 43 - JUL 43	*
WGCDR J.P. RYLAND, DFC	JUL 43 - AUG 43	*
GPCAPT W.H. GARING, CBE, DFC	AUG 43 - MAR 44	
WGCDR J.P. RYLAND, DFC	MAR 44 - DEC 44	
WGCDR G.D. NICOLL, DFC	DEC 44 - APR 45	
GPCAPT G.C. HARTNELL	APR 45 - AUG 45	
GPCAPT K.R.J. PARSONS, DSO, DFC	AUG 45 - FEB 46	
GPCAPT D.W. KINGWELL, DSO	FEB 46 - MAR 48	
GPCAPT A.D. CHARLTON, OBE	MAR 48 - SEP 50	
WGCDR J. DOWLING	SEP 50 - MAR 51	*
WGCDR W.E. TOWNSEND	MAR 51 - FEB 53	
WGCDR C.F. READ, DFC	FEB 53 - JUN 53	*
GPCPAT W.N. GIBSON, DFC	JUN 53 - OCT 54	
WGCDR J.E.S. DENNETT	OCT 54 - FEB 55	*
GPCAPT K.R.J.A. PARSONS, DSO, DFC, AFC	FEB 55 - MAY 58	
GPCAPT E.J. FYFE, DSO	MAY 58 - NOV 60	
WGCDR F.S. ROBEY	NOV 60 - FEB 61	*
GPCAPT A.D. HENDERSON, OBE	FEB 61 - DEC 62	
WGCDR H.C. PLENTY, DFC	DEC 62 - FEB 63	*
GPCAPT W.N. LAMPE, OBE	FEB 63 - DEC 64	
GPCAPT J.G. CORNISH, MVO, AFC	DEC 64 - MAR 66	
WGCDR J.H.D. BLACKWELL	MAR 66 - MAY 66	
GPCAPT J.H.D. BLACKWELL	JUN 66 - JAN 68	
GPCAPT A.R. HODGE, AFC	JAN 68 - AUG 70	
GPCAPT J.S. WILSON, AFC	AUG 70 - JAN 73	
GPCAPT H.K. PARKER	JAN 73 - DEC 73	
GPCAPT K.M. STAIB	JAN 74 - JAN 76	
GPCAPT H.J. HURLEY	JAN 76 - JAN 78	
GPCAPT T.C. OWEN	JAN 78 - MAY 79	
GPCAPT R.E. GILLARD	JUL 79 - DEC 80	
GPCAPT G. DYKE, DFC	JAN 81 - JUL 83	
GPCAPT A.J. PAPPIN, CVO, AM	JUL 83 - JUN 86	
GPCAPT J.D.M. EDWARDS	JUL 86 - JUN 87	
WGCDR B.J. ESPLAND	JUN 87 - AUG 87	*
GPCAPT D.J.S. RIDING, DFC	AUG 87 - DEC 90	
GPCAPT P.J. McDERMOTT, AM	DEC 90 - JAN 93	
GPCAPT J.T. DAY	JAN 93 - DEC 95	
GPCAPT J.D. SMITH	JAN 96 - DEC 97	
GPCAPT P.F. MORAN	JAN 98 - JAN 99	
WGCDR A.J. DANISCHEWSKI, AM	JAN 99 - PRESENT	

* Temporary Commanding

Note: Ranks and decorations are listed at the time members held the position.

ANNEX B

AIRCRAFT OPERATED AT EAST SALE

1930's

Wapiti
Demon

1940's

Battle	Anson	Link Trainer
Hudson	Ventura	Beaufort
Vengeance	Kittyhawk	Oxford
Wirraway	Liberator	Mustang
Tiger Moth		

1950's

Lincoln	Vampire	Winjeel
Dakota		

1960's

U-2 (USAF Det)	Canberra	Jet Provost
Macchi	HS748	

1970's

CT-4

1980's

PC9

1990's

B200

Note: Decade of first appearance is shown. Some types lasted several decades!

Index

prepared by Gary Weston

List Of Illustrations